New Directions in Physics
The Los Alamos 40th Anniversary Volume

New Directions in Physics

The Los Alamos 40th Anniversary Volume

N. Metropolis, D.M. Kerr, Gian-Carlo Rota, editors

ACADEMIC PRESS, INC.
Harcourt Brace Jovanovich, Publishers

Boston San Diego New York
Berkeley London Sydney
Tokyo Toronto

ACADEMIC PRESS, INC.
Orlando, Florida 32887

United Kingdom Edition published by
ACADEMIC PRESS, INC. (LONDON) LTD.
24–28 Oval Road, London NW1 7DX

Library of Congress Cataloging-in-Publication Data

New directions in physics.

1. Physics—Research—United States. 2. Los Alamos
Scientific Laboratory. I. Metropolis, N. (Nicholas),
Date . II. Kerr, Donald M. III. Rota, Gian-Carlo,
Date . IV. Los Alamos Scientific Laboratory.
QC44.N49 1987   530   86-32055
ISBN 0-12-492155-8 (alk. paper)

87 88 89 90   9 8 7 6 5 4 3 2 1
Printed in the United States of America

Dedicated to the memory of

J. Robert Oppenheimer
Enrico Fermi
Niels Bohr
Geoffrey L. Taylor

Contents

Preface

On April 1, 1943, some fifty scientists and engineers gathered on a forlorn mesa in the desert of New Mexico. In the inadequate buildings of a former prep school for boys, they set up a laboratory for the production of a new weapon, what two years later would be called the atomic bomb. It was probably the greatest gathering of intellect of all times, with the possible exception of Athens in ancient Greece.

The explosion of the first atomic weapon was a turning point in history.

For the first time in our civilization, physics played a major role in the future of mankind, a role that lasts to this day. Since that time, no one has seriously doubted that the science of today determines the world of tomorrow.

The Manhattan Project has taught us a lesson in success. No form of interaction will bring scientists together as well as the pursuit of a common goal, one that calls for their best performance and demands the tension of agreement and cooperation. Chemists, physicists, mathematicians, engineers, prima donnas and lab technicians, Nobel Prize-winners and wild-eyed graduate students went on a first-name basis, rolled up their sleeves, and made sure that nothing went wrong. Everyone's contribution was equally essential. A world-renowned physicist loaded with honors would volunteer to monitor the cyclotron during the night, while the experimentalist squeezed in a few hours of much-needed rest.

It was a shotgun marriage that worked. Maybe, in our time, more such shotgun marriages might do us good.

Together with the day-to-day frenzied work on the weapon that was to win World War II, the great postwar ideas of science were being hatched in Los Alamos. The barrier of nonlinearity, the future impact of computers, the astonishing predictions of quantum electrodynamics were being worked upon in conspiratorial whispers between long stints at the bench or in a corner at some loud drinking party on a Saturday night. Los Alamos was to be the birthplace of the new sciences of today as well as of the atomic age.

Those who participated in the effort of those first years have felt to this day the Los Alamos spirit of camaraderie. They have not forgotten the freshness, the wonderment of the two years that for many of them were to be the turning points of their careers and that has joined them closer than any school tie.

No one missed the call to return to Los Alamos for the celebration of the fortieth anniversary of the Laboratory, even when their presence here demanded a major change in their engagement schedule. Their contributions, now collected in this volume, betoken Los Alamos's role as a cradle of scientific talent.

Foreword

In 1983 Los Alamos was revisited by many of the pioneers who established the Los Alamos Laboratory and Project Y of the Manhattan Project. The war years at Los Alamos were ones of remarkable accomplishment, with enormous scientific and military impact.

This book represents a fascinating view of the future as seen by some of the remarkable men who were here over 40 years ago. It makes it quite clear that we are still in the dawn of physics—the excitement and challenge that lie ahead are extraordinary. We also get a glimpse of where these remarkable men have been since the end of the project and where they see the future directions for physics.

The Los Alamos National Laboratory as it is now and some of the prospects for the future are discussed in the lead-off chapter by Donald M. Kerr, Director of the Laboratory at the time of the 40th Anniversary. I concur with Don Kerr that after 40-plus years the Laboratory's prime mission is to continue to help provide the technical basis for this nation's deterrent through nuclear weapon technology. This responsibility remains paramount because deterrence is dynamic and must be maintained to preserve stability and world peace.

I also see the Laboratory having the much broader responsibility of continuing to explore the frontiers of science and keep an eye toward applications for national security. In the long term, national security encompasses not only defense, but also energy security, economic strength, health and environment. We must carefully assess the areas in which the special talents of the Laboratory, such as a strong tradition in basic research, a multidisciplinary approach to science and engineering, and the ability to solve large, complex problems, can make major contributions to national security.

The men who authored the remaining chapters of this book founded the Laboratory with a tradition of excellence. We will strive to live up to that heritage.

Siegfried S. Hecker, Director
Los Alamos National Laboratory

1   Los Alamos in the 1980s

Donald M. Kerr

Nineteen eighty-three marks the fortieth anniversary of the establishment of the Los Alamos Laboratory. And a remarkable beginning it was.

Working in haste in makeshift laboratories with borrowed equipment, a group of dedicated men and women developed the first atomic bomb, the weapon that ended World War II. It is on the foundation of excellence, laid down by those pioneers of the nuclear age, that the Laboratory has grown to become one of the most prestigious scientific institutions in the world.

Its primary mission—to develop technologies for national security—has remained essentially unchanged over four decades. But the scope of its scientific research has diversified.

Los Alamos has entered dimensions of science few imagined in the early days of the Laboratory. Using our outstanding technical facilities and one of the largest computing centers in the world, scientists at Los Alamos investigate an astounding range of phenomena that extends from the earth's interior through its atmosphere and magnetosphere, from subnuclear particles to galaxies, from events occurring in trillionths of a second to those that take thousands of centuries, from temperatures near absolute zero to those measured in tens of millions of degrees, and from the energy of a single photon to that of a supernova. There are few limits.

NEW DIRECTIONS IN PHYSICS
The Los Alamos 40th Anniversary Volume
ISBN 0-12-492155-8

But it all began 40 years ago with those extraordinary people who left a lasting mark on science. Our anniversary is a time to look back with pride on their accomplishments and with appreciation for their legacy. It is a time, too, for looking ahead, for focusing our tremendous resources on the responsibilities and challenges of the coming decades.

My first hope is that Los Alamos scientists will play a prominent role in reshaping the defense posture of America through technical efforts in advanced weapons development and arms control.

The people of this planet have no more important task than to subdue the spiraling arms race and to eliminate the fear that, by accident or by design, nations might eliminate large portions of life on this earth by engaging in a massive nuclear exchange. While science cannot solve the political problems that snarl arms control talks, improved technology in satellite surveillance, seismic detection, and information analysis can help decrease the possibility of treaty violations through surprise actions, clandestine activities, or new developments. Such technological assistance is not likely to be the key element in advancing attempts to curb the arms race but may be useful if political developments become favorable.

Our nation's efforts toward arms control must be made from a position of strength. And that strength depends on being at the forefront of all scientific areas likely to yield new military applications. In the area of advanced weapons development, Los Alamos can make major contributions to:

- Support our nuclear deterrent posture by developing advanced strategic warheads for Trident, Minuteman, and cruise missiles.
- Encourage modernization, where appropriate, of nuclear warheads, to provide the best safety and security features technology can offer.
- Vigorously investigate new approaches which emphasize defensive capabilities rather than assured destruction.
- Apply modern technology to advanced conventional armaments to improve their effectiveness and reduce U.S. dependence on nuclear weapons.

Not only should the purpose for which this nation maintains its deterrent be understood, but also the devastating nature of nuclear warfare should deterrence fail. Effective arms control is an essential element in diminishing the risk of nuclear war while preserving our liberties. However, as we consider the threat of mass destruction, we must consider simultaneously the threat of aggressive totalitarianism. Both are central to the political, military and economic dilemmas of our age—it is naive, false, and dangerous to assume that either of these can be ignored and the other dealt with in isolation. We should always be alert for opportunities for technology to allow us to deal with the intertwined problem of assuring survival in a nuclear armed world and our geopolitical struggle with the Soviet Union.

My second hope is that the Laboratory will make major contributions to solving a problem that has commanded great public attention—the problem of supplying the future energy needs of the nation and the world. The Laboratory has devoted a substantial effort to energy programs during the past decade, and it is my hope that as these efforts reach maturity in the coming decade, they will bear technological fruit in the following forms:

- Safety and engineering advances that will make nuclear power a more acceptable approach when the world turns again to this energy source, as I believe it eventually will.
- Nuclear waste disposal techniques that will satisfy public concerns.
- Techniques for extracting fossil fuels from the earth that will provide greater efficiency and worker safety and cause less pollution and environmental damage.
- Advances in renewable energy technologies (e.g., solar and geothermal) that will allow for decentralized energy supplies so necessary in rural America and in many developing nations.

Controlled fusion is a major area in which we now make important contributions to the development of a new energy source for future use, and we can continue to do so. Since the early 1950s, Los Alamos has played a major role in the international development of magnetic confinement science and technology. This cooperative effort has led to such a high level of sophistication that demonstration of energy break-even, using the mainline Tokamak approach, seems assured during this decade. The ability to confine reactor-grade plasmas for times close to those required for thermonuclear ignition is an enormous scientific accomplishment which could not have been achieved without the resources that national laboratories, universities, and industry brought to bear on this problem.

My third hope concerns the application of the Laboratory's expertise in physics, chemistry, and engineering to the new challenges in the fields of biology and medicine. Two instruments of fundamental importance to biomedical research have been developed at Los Alamos. These are the liquid scintillation spectrometer, which makes possible simultaneous counting of different radioisotopes, and the flow cytophotometer, which allows rapid analysis and isolation of individual cells. The latter development resulted in the establishment at Los Alamos of the National Flow Cytometry Resource. Current activities give me confidence that the next decades will see developments of similar importance to biology and medicine.

For instance, we are involved in the development of noninvasive techniques for analyzing human functions with minimal discomfort to the patient. In one such technique nuclear magnetic resonance (NMR) is used to follow the course of metabolic processes from outside a patient's body. The NMR technique detects important intermediate products of metabolism that have been labeled with a suitable isotope, such as carbon-13. The labeled materials are made possible by our pioneering work in producing stable isotopes for biomedical research.

Another venture into the realm of biology exploits our scientific computing capability—the largest in the world—to compile and make available to the scientific community a library of genetic sequences. Los Alamos has recently been designated as the site of The National DNA Sequence Data Bank. This data bank will contribute significantly to unraveling the mysteries of DNA.

My fourth hope is that the Laboratory will continue to involve an increasing number of scientists from universities and industry in its activities. We have already made great progress in this area by establishing three centers designed to reach aggressively beyond our borders: a branch of the University of California Institute of Geophysics and Planetary Physics, the Center for Nonlinear Studies, and the Center for Material Sciences.

It is, of course, impossible to mention all significant advances expected in a laboratory as diverse as Los Alamos. But one final hope is that we will be surprised by some unexpected development or discovery that derives from the exploration of new questions and new possibilities. The very nature of scientific research makes such surprises possible, and for this reason basic research is a fundamental element in our plans.

To realize the hopes that I have outlined, difficult scientific problems will have to be confronted, pursued, and conquered. Beyond the inherent scientific difficulty, a changed political and social climate now challenges these hopes. Some voices now question the major mission of the Laboratory. They ask, "Why is the Laboratory still engaged in weapons work?" That question often comes from those who believe that the thousands of nuclear warheads now in our arsenal are more than adequate and that no more effort in this technical area is needed. These people deserve a reply.

Three chief factors drive our continued efforts in weapons. The first is the extent to which potential enemies of the United States are making technological advances that could jeopardize the defense posture of the

United States. This issue led to the creation of the Manhattan Project during World War II, and it is still a valid concern in the present political climate. Our political leaders generally feel that their ability to influence world affairs is affected by the extent to which the United States maintains technological supremacy in the defense area.

The second factor is the need for solutions to technical problems that may inhibit accords on arms control. Any agreement on this subject rests heavily on the ability to determine that its provisions will be followed by each signatory. The inability to verify compliance has created stumbling blocks in past negotiations. The Laboratory must assist in developing new verification techniques, for they may be a critical link in reaching the goal of arms control. The Laboratory will also be called upon to help policymakers understand the capabilities and limitations of current approaches to verification.

The third factor is the certain knowledge that the pursuit of science inevitably yields ideas for new technologies that have a wide variety of applications, including military ones. The choice to develop the new military applications is the nation's. But the nation cannot choose to stop the scientific effort that creates those applications without also stifling development in other human endeavors. Science is neither compartmentalized within itself nor isolated from its surroundings. New scientific ideas have a way of leaping traditional boundaries among fields of science and of creating vast and unforeseen changes in the economic and political fabric of society.

Another challenge facing the Laboratory is the question whether some of our research activities should be transferred to academia and industry. You might ask, "What is the place of Los Alamos in the midst of the country's large and sprawling research community?" After all, research efforts at universities have grown substantially since World War II and industry has also seen reason to invest in research and development.

I believe there is a clear place for Los Alamos and other national laboratories. That place goes beyond weapons work, which the government obviously must control directly, to other areas of research in which a strong national interest justifies the presence of a federally supported laboratory.

For example, many areas of research—a notable example being nuclear fusion—face such inherent difficulties that they will yield results only over a very long term. Industry will neither be inclined nor financially able to enter such areas. Another example is the area of research on the protection of workers, the public, and the environment from technologies new or old. Here the profit motive of industry may bring into question their objective assessment.

National laboratories such as Los Alamos can address these issues, and, in fact, Los Alamos is extraordinarily well equipped to do so. Our scientific computing capabilities are unsurpassed. We have the experience of dealing with military agencies, and understand their needs and procedures. We can work in a way sometimes referred to as vertical integration: that is, we can develop an idea for, say, an instrument all the way from conception to production engineering. Our activities range from undirected basic research to production engineering of devices that weigh tons. We can transform ideas or bits of Nature's secrets into products useful to mankind. Of the thousands of laboratories in the nation only a small handful approach the Laboratory's capabilities.

The world is increasingly specialized, compartmentalized, separated into isolated parts. The concept of integrated teamwork bringing mathematicians, physicists, chemists, biologists, engineers, and economists together for a sustained effort is not a tradition at very many institutions. In fact, it seldom happens. It is difficult to bring about. In many places it is impossible. At Los Alamos it is the usual practice. It is the way we have conducted business from the beginning.

Let me conclude with a final challenge—the desire of some that science should overcome the tangled web of politics and assure that all its results are used only in positive ways. Such a desire is natural, but it is too much to expect of any single sector of society.

At the end of World War II, those at Los Alamos learned with the rest of the world that technical developments were beyond the control of the small group of scientists who pleaded that the results of their work be used solely for peaceful purposes. That control rests with the broader institutions of society. Today we continue to pursue the unanswered questions of science in the belief that our efforts will enhance the peace and prosperity of the world. The ultimate hope of those of us at Los Alamos is that the voices for peace will prevail in all decisions that affect the use of our endeavors.

2 Tiny Computers Obeying Quantum Mechanical Laws

Richard P. Feynman

Although the computing business has changed from the alphanumeric type bar zone entry to the MOS VLSI, they continue always to hide the computer behind those crazy words. As I found out, it isn't as complicated as the words make it sound and I've had an interest in computers and computing from the early days. So although I have done mostly physics, I from time to time pay attention to computers.

Two years ago Carver Mead, a professor of computer science at Caltech, discussed with us some of the engineering and design problems of computers. He felt there ought to be physical laws about the limits in computer design. For example, one design problem is the amount of heat generated by an operating computer. This heat is rather large and you have to get it out. For many purposes, it's a good idea to make computers small because then the time delay for signals to go from one part of the computer to another is much less and the computers can run faster, but if you make them small, it's hard to get the heat out. Carver felt there must be a theorem that gave the minimum amount of heat you could generate.

I got interested in the problem and I worked it all out. It turned out that Charlie Bennett from IBM had worked it all out five years earlier. It's just that I wasn't in the business, and Carver didn't happen to believe Charlie. He wanted an independent demonstration.

7

NEW DIRECTIONS IN PHYSICS
The Los Alamos 40th Anniversary Volume
ISBN 0-12-492155-8

A revised version of this article appeared in
Optics News 11(2):11–20, Feb. 1985, under the title
"Quantum Mechanical Computers."

The next question was what are the limits in computers due to quantum mechanics? If we make the computers very small indeed, then the laws of physics are no longer classical but quantum mechanical. What I hoped to do was to design a computer in which I knew how every part worked with everything completely specified down to the atomic level. In other words I wanted to write a Hamiltonian for a system that could make a calculation. Then, in terms of that Hamiltonian, which is a method of stating the physical situation completely, I could calculate the various effects of the limits due to quantum mechanics.

I will summarize my conclusions at this point by stating that there are virtually no limits due to quantum mechanics except for size. You can't get the numbers any smaller than some atomic unit because you've got to have something to write on. However, the thermodynamic limit determined by Bennett also applies to the quantum mechanical system, and the quantum mechanical system presents no extra fundamental limit, at least with regard to the consumption of energy.

The first physical limit of computers that I'll concentrate on is the heat or energy problem. The size problem is obvious, and there is very little to say, except that anyone who builds computers soon discovers that a fundamental difficulty is the wire space. Most theorists don't notice that wire space is a problem. For them a wire is an idealized thin string that doesn't take up any space, but real computer designers soon discover that they can't get enough wires in. This is not, however, a fundamental but a temporary limit resulting from the two-dimensional present design of all computer circuitry: everything is on the surface of the silicon. If it could be given a third dimension by using the depth, we would have enough directions to put the wires in. It's like traffic jams on the street that are offset by airplanes going up and down. You've got another dimension. Traffic jams that come in wiring are only a temporary difficulty, and I will leave that out, though I know it's an important practical limit at present.

Thermodynamic Limits

You know of course that all computers, no matter how complicated, are made of primitive elements in great multitude. The number of these very simple circuits may be half a dozen or even just two or three. Figure 2.1 gives an example of four primitive elements out of which all kinds of computers can be made by putting these four elements together in different ways with a lot of wire.

The system of representing numbers that is most convenient in computers, of course, is base two in which there are just two values, zero

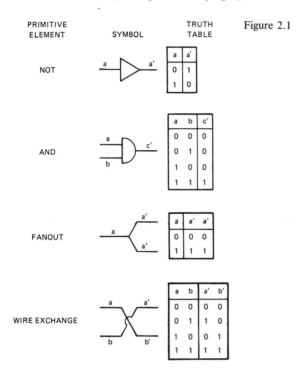

Figure 2.1

or one, for any bit. All the numbers are just strings of such zeros and ones, which, in electric devices, are usually represented by voltages on a wire. If the voltage is up, it's a one, if down, it's a zero.

The first primitive element shown in Figure 2.1 is called "not". A wire comes in, goes through the element, and goes out the other side. If the input wire has a voltage on it then the device turns the voltage off on the output wire. If the input doesn't have a voltage, the device puts one on the output. In other words, if you think of zero and one as false and true, this element changes the output to its opposite: what was false becomes true, what was true becomes false. Figure 2.1 displays this truth table next to the element.

Another primitive element is "and". Here there are two input wires, a and b, and one output wire, c'. (Why did I put the prime on c? Lots of good questions in this field.) For this element, if a and b are both zero then c' is zero. If only a or only b has a voltage of one, then still nothing happens. The only way you can get a nonzero output on c' is to have both voltages on a and b equal one. Of course, we have to make a circuit to do that, but "and" is an example of a primitive element.

Two elements, which are usually not considered but have to be in my type of fundamental treatment, are "fanout" and "wire exchange". "Fanout" is just a wire branch. It's a simple thing to build, but when you address fundamental questions you have to be aware that you are branching. A "fanout" has two output wires that always agree with the voltage on the single input wire.

A "wire exchange" is an even crazier thing and just a matter of wiring, but for my purposes I have to be able to handle what corresponds logically to "wire exchange". You have two wires and the information gets switched from one to the other. The wires coming in are called *a* and *b*, but coming out are called *a'* and *b'*. If *a* and *b* are both zero, so are *a'* and *b'*. If *a* is zero and *b* is one, then the *a'* gets the one and the *b'* gets the zero. In other words I simply exchange voltages.

Those four elements are enough to build anything else. But you ask, what about the various kinds of logical operations? They can all be made out of these elements so that, in principle, all we have to discuss is these four elements. You could choose other ones for efficiency and so on, but I am here dealing with matters of principle and, at this level, don't want to discuss engineering efficiency.

Today these elements are made from transistors (Fig. 2.2). A transistor amounts to a thing that is just a switch controlled by electric voltage. Part a of this figure shows how this works. If there's no voltage on the gate, then resistance is very high (the switch is open). If there's a positive voltage, then resistance is low and current flows through the transistor to ground. This device resembles the primitive element "not". If there's no input voltage on the gate at *a*, the high resistance in the transistor

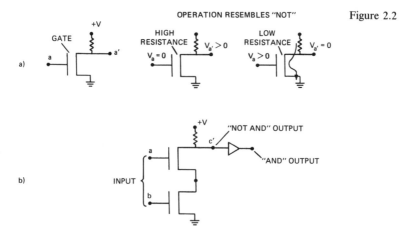

Figure 2.2

means that a nonzero output voltage builds up at a'. On the other hand, if I do have a nonzero input voltage, the semiconductor is a very good conductor and the output voltage goes to ground or zero. So you see that when one is up the other is down and vice versa.

To make an "and" element we put two transistors in succession along the wire as shown in Figure 2.2b. If either or both of the input voltages (a and b) are zero, then a nonzero, positive voltage will build up at the output wire c'. Only when *both* gates are open, with two positive input voltages, can the output voltage drop to zero. The truth table for this arrangement of transistors resembles the inverse of "and", that is, "and" with the wrong sign. So I put a "not" element on the output to turn it over. This final circuit gives a plus voltage on the output when the two input voltages are both positive, but otherwise the output is zero.

The "fanout" is easy, just split the wires. Likewise for the "wire exchange", just cross the wires. And that's the way things are made today—with thousands and millions of these elements on little chips all wired together.

Now I want to discuss Bennett's analysis of the heat limits for such devices—not actual devices, but in principle. Think about the primitive "and" element. In effect, this device makes a decision by finding out whether or not the input voltages are the same. When it produces an output, a decision with regard to a and b has been made. Now you may guess that when this happens for any primitive operation the entropy changes by an amount equal to ln 2. You guess this because you are flipping a wire between zero and one; you are making a decision between two possibilities. Thus, the energy needed by the element to operate at a certain temperature should be of the order kT. It's not hard to come to that conclusion. I won't belabor it much because it's wrong.

Actual devices at present made from the transistors in Figure 2.2 use about 10^{10} kT instead of kT in making that decision. It's obvious that this is just gilding the lily. The right thing to do is to redesign so that the energy use per bit drops from 10^{10} kT to, say, 10^5 kT, and you would solve all the practical problems. In principle, if this theory were right, you could even get a 1 kT energy use, but, as I will show, it is very much less than 1 kT. We don't have to get less, because if we get down to 1 kT we have already improved things by a factor of 10^{10}. So there are good engineering problems here. But I love principle, and I keep right on going anyway even though there's plenty of clearance between what we do today and what was once thought to be the limit but which Bennett proved is not.

Bennett also noticed that DNA, for example, is a kind of copying machine. It takes a number—the information consisting of bits on DNA—

and makes another copy. It's something like a computer. (In fact, Bennett showed how a copying machine can be made into a computer.) DNA is a primitive element that permits you to make a computer using only 10^2 kT of energy per bit. Nature as usual is a hundred million times more efficient than man at the present time, but nature is still a hundred times less efficient than the old theory which turns out to be a hundred times less efficient than you could, in fact, make it.

It turned out that what was wrong was considering primitive elements to be irreversible. Everybody thought that a computer must be irreversible. Bennett and, independently, Fredkin and Toffoli have analyzed the idea of a reversible computer and shown that it's something like Carnot heat engines in which the reversible one is the efficient one. So the irreversibility that everybody thought was implied in the primitive elements was un-necessary. They were badly designed.

One of my primitive elements, the "and" gate, is irreversible. There are three different states that give the same zero output. If you knew only this output you couldn't go back and figure out which of those three input sides it was.

The "not" element, however, is reversible. If I perform the "not" operation twice I change zero to one then back to zero again, in effect, leaving it alone. So "not" is an example of a reversible primitive operation.

Now Bennett first discovered that if you have a reversible machine the minimum energy requirement is essentially zero. To put it another way, you only need kT per bit of the answer no matter how complicated the calculation is. You can have millions and millions of primitive elements doing the calculation, but if the answer has only 40 bits then 40 kT is the minimum energy needed. As a matter of fact, even that amount turns out to be just the entropy needed to clear the output of the computer back to zero so it can be used again. You have to think in a cycle. If you realize that clearing the computer means taking the answer away to do something with it, you might as well associate that entropy with what you *do* with the answer rather than with the computer that generated it. It's not very important, because it's a very small amount. Because it doesn't depend on how complicated the calculation is, but only on how many bits it takes to write the answer—if it depends on that at all—I'd like to say it takes essentially no free energy. The limit is zero. Of course, this is only an ideal limit. If you compute with a perfectly reversible computer, you also compute at infinitesimal speed. It's just like Carnot's cycle. (Carnot would say in a lecture "I've got an idea for the efficiency of a heat engine, but you have to make it reversible. Of course, to make it reversible you have to work infinitesimally slowly." Everybody in the audience would laugh because it can't go. Nevertheless, that's the ideal limit.) What is analyzed next are the effects due to friction and various things: how much heat is actually needed to get going a certain speed and so forth. The principle with computers is very much like the principle of Carnot.

Let us look at the reversible computer. An exciting discovery, made mostly by Fredkin, was that you can make a computer solely out of reversible primitive elements. There are many ways to do this, but, for simplicity, I'll show only one. There are other reversible primitive elements that, for certain purposes, are a little easier. But I'm going to do it all with one primitive element, the Fredkin gate (Fig. 2.3). Three wires come in and three wires go out. Line *a*, the top wire, is considered a control. It comes in and goes out the same; nothing happens to it. The other two wires, *b* and *c*, are the ones being controlled. What happens is that sometimes their voltages or information is exchanged. If the *a* wire has a zero on it, nothing happens; both *b* and *c* voltages go through the element unchanged. But if the *a* voltage is positive, or one, then the *b* and *c* wires exchange their information. The truth table in Figure 2.3 gives a complete list of the possible states that can come in on the three wires and then what will go out on the other three wires. The only two cases with an effective exchange are when the *a* wire has a one and the *b* and *c* wires differ from each other. If the *a* and *b* voltages are identical, the exchange is without apparent effect.

I'll now show that the Fredkin gate will produce everything we need.

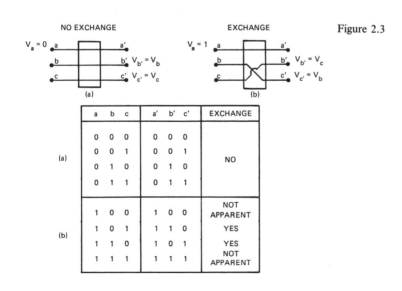

Figure 2.3

However, in addition to the data or problem input, we have to supply some wires on which there are ones and zeros available for control, that is, for putting the gates in the proper state for the calculation. Let's examine the Fredkin gate for the case in which b has a one and c has a zero (Fig. 2.3). What comes out? Well, along the a wire the voltage stays the same: a comes out a. But if a equals one, there's exchange on wires b and c, and c ends up with one, that is, equal to a. If a equals zero, there's no exchange and c ends up with zero, or, once again, equal to a. Thus, by putting one and zero on b and c respectively, the a' and c' output wires act essentially as a fan out for wire a.

What about the b' output? If a is one, b', because of the exchange, becomes zero. If a is zero, b', without exchange, stays one. This is equivalent to "not a". So by putting one and zero on wires b and c, respectively, we generate two of the necessary primitive elements.

The "and" element is also easy. To make it you put a zero on line c, treat a and b as the input and c' as the output. For the case that a equals zero, there's no exchange and c' stays zero. If a equals one and b is zero, there's exchange, but c' still ends up zero. Only in the case that both a and b are one do you get exchange resulting in a one at c'. This is the required behavior for "and". In addition, the b wire has voltages consistent with "(not a) and b", but we don't need this. "Wire exchange", of course, is produced by putting one on the a wire, then treating b and c as input and b' and c' as output.

With one primitive element we produce all the effects we need. In addition, the Fredkin gate is reversible, as can be seen by what happens if you use it twice. If a has nothing on it, things go straight through even if you do it twice. So you're still in the initial state. If a has a one on it, everything is exchanged. Put another Fredkin gate there with a equal to one and you get a second exchange. Two exchanges bring things back to their original state. Therefore the Fredkin gate is a reversible device. In fact, if you reverse the columns of its truth table, you'll find the same corresponding states for input and output as before, just in a different order. Because this is a reversible element, reversible computation is possible. This was a curious feature when it was discovered.

In actual practice you have to put a lot of these elements together. Imagine a whole bunch of wires representing the digits of some number that you want to manipulate. You put together a series of Fredkin gates with "and" and "exchanges" and so forth to do the manipulating. In this way you make a logic unit. I won't bother you with the details (even though they're very interesting) about how you can put these logic units together and get branching effects that send you one way if a certain wire has a one on it and send you the other way if it's got a zero. It's the standard way of putting things together, almost, but for reversible computers you have to be a little more careful.

For those of you who are familiar with this or else are astute, you'll realize that, in addition to the answer, a lot of garbage comes out. There are wires we don't want. Furthermore, we have to put more in than just the input number; we have to use lots of zeros and ones to make the system operate right. What about this garbage?

Figure 2.4 symbolizes a large system of logic units and branches and what else. The input has a number for the data in the problem and a number that's only a bunch of zeros, the standard zeros, that are used for control. You ask, what about the ones needed for control? Put a "not" (in addition to the Fredkin gates) after appropriate zeros and you've got your ones.

Now the answer comes out the other side onto a register, but there are also a lot of other wires with junk we don't want like "(not a) and

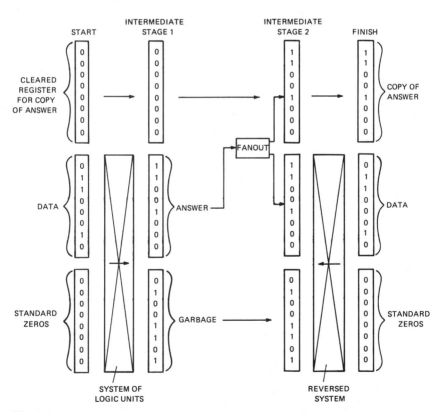

Figure 2.4

b" and so forth. This garbage represents a kind of heat, a kind of entropy. So when you're counting the entropy have you fooled yourself? No, because, as shown in the figure, what you can do is copy the answer by fanning it out into two registers each of the same length as the answer but initially set with zeros. So we have two copies of the answer—a dumb thing to do. Who needs another copy of the answer? Well, we take the same computer but with everything reversed, everything built in the opposite order, and put one copy of the answer and all the garbage from the first computer through the reversed computer. What happens? You get back the original data and the standard zeros. All the garbage, entropy, has disappeared. In fact, because you get back the zeros you never have to talk about them; they can be stored inside the computer. In effect what you do is double the equipment and make it take twice as long to do a calculation, but there's no entropy loss. You start with an input, put zeros in the copy register, and at the end of the calculation that register contains the answer and an input register contains the input. No losses. Nice and reversible. There's no problem about the entropy and the garbage.

Quantum Mechanical Limits

Where I made a contribution, I think, is with quantum mechanical limits. I'll probably discover by looking at the literature that I'm only three years late with this one.

I've already said you can't make a computer element much smaller than an atom or some simple thing to write a number on. How do you write a number on an atom? All we need is something like the wire which can be in two states. Very simple. Say we have some kind of quantum mechanical system with two states so that if the system is in the up or excited state, we call it a one, but if it's in the down or unexcited state, we call it a zero. (Here I'm not going to worry about other possible states of the system.) I'll call such a two-state thing an atom. It could be an electron, it could be something smaller. I'm not going to be involved with the actual design or arrangement of atoms, but I'm going to show that you can write down a Hamiltonian that represents couplings between these various atoms. Now, we can, in principle, make a computing device in which the numbers are represented by a row of atoms with each atom in either of the two states. That's our input. The Hamiltonian starts "Hamiltonianizing" the wave function, that is, the state of the row of atoms changes as a function of time according to the dynamics of quantum mechanics. The ones move around, the zeros move around, other atoms interact. Finally, along a particular bunch of atoms, ones and zeros or excited and unexcited states occur that represent the answer. Nothing could be made smaller than that. Nothing could be more elegant than such a device completely described in every atomic motion. No losses, no uncertainties, no averaging, no nothing. But can we do it?

Let's see what a Fredkin gate requires. First, it has to have at least three atoms because there must be a two-state atom that corresponds to each two-state wire. A three-atom system then could assume eight states corresponding to the states in the Fredkin-gate truth table (Fig. 2.5). Although there are other linear combinations, I prefer eight base states. As a matter of fact, the logic is easiest if these states are the base states and if we say that the system will be in only one or another of these base states at any time. Now, the original Fredkin gate operated by taking voltages from one set of wires and placing new voltages on a second set of wires. Here I let the Fredkin gate smash onto my three atoms, but when it retires from the field the new state will be on the same three atoms, not on a new set of atoms. Actually, it doesn't make any difference, it's just a little easier to think about.

What I've said is represented mathematically in Figure 2.5 by an equation with a Fredkin-gate operator F, an eight by eight matrix that operates on our three-atom state (a, b, c) to convert it to the appropriate state (a', b', c'). I'd like to point out that because the Fredkin gate is reversible, F, if applied twice, brings you back to the state (a, b, c). In other words, $F^2 = 1$. In general, I want F times its complex conjugate to equal one because that represents reversibility in quantum mechanics. But my matrices will be real. So instead I'll have $F = F^*$ and $F^2 = 1$. In fact, I'll assume reversibility for all couplings because the microscopic laws of quantum mechanics are reversible. As a result I can't invent

THREE-ATOM STATES   Figure 2.5

INITIAL			FINAL		
a	b	c	a'	b'	c'
0	0	0	0	0	0
0	0	1	0	0	1
0	1	0	0	1	0
0	1	1	0	1	1
1	0	0	1	0	0
1	1	0	1	0	1
1	0	1	1	1	0
1	1	1	1	1	1

$$\begin{pmatrix} a' \\ b' \\ c' \end{pmatrix} = F \begin{pmatrix} a \\ b \\ c \end{pmatrix}$$

a = INITIAL STATE OF ATOM a, etc.
a' = FINAL STATE OF ATOM a, etc.

F = FREDKIN-GATE MATRIX OPERATOR

something like the "and" element at the quantum mechanical level because "and" is an irreversible element.

Now let me give examples of matrix operators that correspond to different kinds of primitive elements. First, the "not" element has a matrix that operates on a single state so that when the state is up, the matrix turns it down, when the state is down, the matrix turns it up. If you think of the two states zero and one as the $-\frac{1}{2}$ and $+\frac{1}{2}$ spin states, you know this matrix as one of the Pauli spin matrices, usually called σ_x and written:

$$\sigma_x = \begin{pmatrix} 0 & 1 \\ 1 & 0 \end{pmatrix}.$$

This operator represents a physical quantity in atomic systems, coupling with the x component of the spin.

Another example is the operator for the "wire exchange" element. This one can also be expressed in terms of σ matrices, one operating only on atom a, another operating only on atom b. In equation form this operator is recognizable as Dirac's famous spin-exchange operator: $P_{ab} = \frac{1}{2}(1 + \sigma_a \cdot \sigma_b)$. Its operation in the two-atom system causes the states of atoms a and b to be exchanged in similar fashion to voltages in the "wire exchange".

We can now talk about the controlled exchange of the Fredkin gate. For that we need a three-atom system in which the state of atom a determines whether or not the states of atoms b and c are exchanged. An operator for such a system is given by:

$$F_{abc} = \frac{1}{2}[1 + (\sigma_z)_a]P_{bc} + \frac{1}{2}[1 - (\sigma_z)_a],$$

where $(\sigma_z)_a = -1$ or 1 for the zero and one states, respectively, of atom a, and P_{ab} is the exchange operator. Thus, if atom a is in the zero state, $F_{abc} = 1$, and nothing happens. However, if atom a is in the one state, $F_{abc} = P_{bc}$, and there are exchanges of states between atoms b and c. We see that F_{abc} can be made up of primitive matrices: P_{bc} is a multiplication of σ matrices from atoms b and c and $(\sigma_z)_a$ is one of the σ matrices for atom a. The main point is that F_{abc} has only matrices involving the couplings between three atoms at a time.

Now we get to the question of the logic unit. We could make this with a large number of atoms that act in sequence in groups of three as Fredkin gates. The resulting calculation would involve a lot of operations, which, if there were n atoms, could be represented by a great big matrix of size 2^n by 2^n. This monstrous matrix, M, is a product of successive operations and thus can be written as a product of matrices, $M = \ldots F_{ijk} \ldots F_{cbd}F_{dab}F_{gce}$. Here each F_{ijk} matrix is again 2^n by 2^n, but, except for the $i, j,$ and k entries corresponding to the three atoms in that particular

Fredkin gate, the matrix has ones along its diagonal. Thus, it involves couplings between those three atoms and not others.

So I have to multiply a long sequence of unitary matrices to find the final result. The question is how can I make the dynamics of quantum mechanics generate such a product of matrices? The matrices themselves would be easy to make with particular special couplings, but the serious problem is how to multiply them together. It has been suggested that this operation, involving unitary matrices, can be represented as the action of some Hamiltonian for a definite amount of time because that is also a unitary operation. Then you would have to let the first Hamiltonian run for a definite amount of time to get the next one, change the Hamiltonian, run it for another definite time, and so forth. That's an awful lot of external machinery, and you don't see what you're doing. As a matter of fact, it's extremely delicate. If you left it running a little too long, the matrix would be wrong. I don't want to discuss something where I have to push buttons from the outside because, for me, pushing buttons is just more atoms doing things. Let's get all the atoms into the system. So I'm going to generate my hypothetical logic unit in a straightforward way directly by atomic couplings. Figure 2.6 shows the n atoms of the logic unit on which I want the F_{ijk} matrices to operate in succession. (To simplify, I relabel these matrices as A_1, A_2, ..., A_i, ..., A_p using one index that increases with the order of the operations to the last or p^{th} matrix.)

To this system I add another set of $p + 1$ atoms that I call the program counter atoms. When these atoms are lined up and numbered from 0 to p, the number of gaps equals the number of operations that need to be performed. If all the counter atoms are in their zero state, nothing happens. Whatever state exists in the n-atom register stays there. Now say I put the counter atom labeled zero into the one state but leave all the rest in their zero states. Then, as time goes on, there's an amplitude for the one to jump to the next counter position bringing the original atom back to zero. Say that when that happens—and only then—an A_1 operation occurs in the n-atom register, that is, the n atoms representing the number undergo the first step of the calculation. In fact, in my proposed system the only way for the one state to move from the counter atom labeled zero to the atom labeled one is for the system of n atoms to change in accordance with the A_1 operator. Next, when this positive counter state goes from atom 1 to atom 2, the A_2 operation gets performed and the n-atom register changes again, and so forth. If I could figure out some way to make this state go straight down the line of counter atoms, I would, in effect, simply multiply the number in the n-atom register by one operation after another.

Also in Figure 2.6, I show the Hamiltonian which describes this idea.

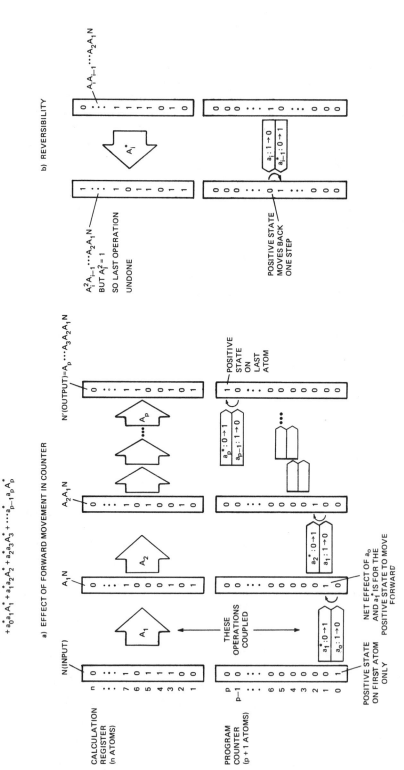

Figure 2.6

It's a little bit wild notation if you're not used to it, but it's easy to explain—just like alpha-numeric type bar zone entry. Each term in the Hamiltonian, $a_i^* a_{i-1} A_i$, corresponds to one step in the calculation, and, in fact, includes the operator A_i for the ith step. The a_{i-1} is a little operation that takes a one on counter atom $i - 1$ and turns it to zero. If there's a zero on that atom, the factor does nothing. The a_i^* factor does the opposite except to the next counter atom. If there's a zero present on atom i, this factor kicks it up to one. These simple factors can also be written in terms of σ matrices:

$$a = \begin{pmatrix} 0 & 1 \\ 0 & 0 \end{pmatrix} = \tfrac{1}{2}(\sigma_x + i\sigma_y).$$

Now the net effect of one of these terms in the Hamiltonian is to cause a one state on atom $i - 1$ (if such a state exists) to jump to atom i, and, at the same time, to cause A_i to operate on the n-atom register. The Hamiltonian, of course, must be what they call self-adjoint, or Hermetian, so you have to have all the complex conjugate terms as well. But the net effect of the term $a_i a_{i-1}^* A_i^*$ is to cause the one state to jump backward from atom i to atom $i - 1$ and to cause the ith operation to happen again to the n-atom system. (This is because the A_i's are real and $A_i = A_i^*$.)

What happens if you let this system evolve in time after putting the first counter atom in state one? The Hamiltonian has a certain amplitude per second of doing what's represented by each of its terms. In other words, every second there's an amplitude that the one state will jump across a given gap in the counter atoms. So the state moves one way or another. Each time it does it, in effect, multiplies the n-atom system by the A_i matrix associated with that jump. If it has moved directly from the zero to the jth counter atom, then the operation represented by $A_1 A_2 A_3$ ... A_j has been accomplished. But what if it now goes backwards one step? Another A_j is multiplied into the product and we have $A_1 A_2 A_3$ ... $A_{j-1} A_j^2$. Because these operations are reversible, $A_j^2 = 1$ and our calculation to that point is now represented by $A_1 A_2 A_3$ ... A_{j-1}. In other words, if the state decides to go backwards, it undoes what it did before. No matter how the state wanders through the counter atoms, when it arrives at a particular atom the multiplication of the matrices has been accomplished up to that point. Therefore, what we do is start with a one state at the first counter position and a zero everywhere else. Then we sit at the last atom and keep watching. When this atom acquires a one state, we take it off so it can't go back. We have finished the calculation. We know the answer on the n-atom register is the product of all the A_i's.

In this way you can put together the fundamental coupling for a logic unit. The A_i operation requires three atoms to be coupled together; the counter operation for that step requires two more atoms to be coupled to A_i; so I now have a five-atom coupling. Such a device is a little hard to design and build in practice but, nevertheless, buildable in principle. There are no particular fundamental problems, and, therefore, we have a Hamiltonian which does make the calculation.

Of course, in a real computer you have very long chains, and branches, and so forth, but I won't go into the details. It's just the idea that, in principle, we have a way in which a quantum mechanical device that computes could be built.

Further Implications

The fact that you have to multiply by the A_i matrix when the ith jump occurs in the counter may make you think that somehow the strength of the transition will be effected. It isn't. It turns out that the motion of the one state along the program counter is exactly the same whether you include the A_i matrices or you have ones in their place. In other words, it's just the same as an up spin riding back and forth on a one-dimensional, coupled-spin system which is sometimes called a spin wave. It acts exactly like that; there is no noise generated by the different matrices. In fact, you can make the whole thing work by ballistic action, which we know a lot about. We could have had a few more atoms before the start of the program counter and many other atoms behind it. Then we start the initial one state with momentum, and it coasts like a wave right through the whole machine. When it goes through the part from the zero to p, it does the calculation on the n-atom register. When it comes out the other end it coasts further to the end of the wire. I don't know where you're going to put it, but you read it or dump it or whatever. This ballistical device, in principle, can make the calculation with no losses whatever and at a high speed.

Fredkin was the first to show how to make a ballistic computer. The reversible computers of Bennett were all thermodynamic diffusive computers but reversible. Fredkin showed that, in principle, if you could have perfect, classical machinery you could make something that would simply bounce across dynamically by collisions. It was called the billiard ball model of a computer. This is another ballistic device.

How might you capture the one state when it arrives at p? There are many ways. You might have a device that only scatters light when the p atom is in the one state. You could also have the state spread from p into a large number of possible states so that once it gets out it can't find its way back. That's an irreversible process.

Because this computer requires the use of five-atom interaction terms in the Hamiltonian, it might be a little hard to actually build. Whether or not it could be built using real electron spins with exactly these couplings, or with little superconducting circuits, or with resonating cavities of very small size with photons in them, or what, I don't know yet. But the use of five-atom couplings is unnecessary. It's possible to do it with only three-atom couplings.

What happens if there are imperfections—not imperfections in the n-atom register, but, just for fun, imperfections along the program counter? In a real computer that program counter is going to be very long, because there's an enormous number of primitive steps in making a calculation— millions of them. If you tried to make a ballistic counter with millions of atoms there would, in fact, be slight imperfections that would cause scattering back and forth. The state wouldn't go through straight. What you'd have to do then would be to apply a small external force to drag it through. Because the force is diffuse, the state would go back and forth a bit and waste a little calculating time, but you'd get it through. Now it's very easy to analyze for the power you have to apply to get the state through the counter at a certain speed. This depends, of course, on how accurately you've made the chain, that is, on the probability of getting a scattering for each primitive step of the calculation. Suppose the device is made reasonably well and q is the probability that a scattering occurs in one calculational step. I found out that if you pull the program atom through the system with external forces the amount of entropy generated per net calculational step, S, is:

$$S = q\ln(f/b),$$

where f equals the probability of forward movement and b equals the probability of backward movement. You could probably understand this equation if I'd said the entropy cost per scattering is $\ln(f/b)$ because that quantity *is* an entropy. When I get a state that was moving forward to go backward instead, I have lost that entropy and I have to supply it. The q factor enters the equation because I want to calculate the cost per calculational step, not per scattering.

Of course, if the probability of forward movement is exactly the same as the probability of backward movement, you have the delightful result that there's no entropy generation, no heat generation whatever. The only trouble, of course, is that if the probability of forward movement is the same as for backward movement, the state will never get through except by diffusing through accidently. But accidental diffusion would take too long for millions or billions of steps. Thus, when we talk about, effectively, an infinitely long computer, we can't wait for the state to diffuse through. We have to pull it so that it's moving forward more of the time than it's moving backward. Now if you pull gracefully—not too hard, so the difference is small—then this logarithm is approximately equal to the probability of forward movement minus the probability of backward. As a result, we can write:

$$S = q(f - b).$$

The difference, $f - b$, is a measure of how fast the state is going through. As a matter of fact, this same formula can be written using Δt_{min}, the minimum time to do the calculation if the state always went forward by good luck. The equation is:

$$S = q(\Delta t_{min}/\Delta t),$$

where Δt is the actual time used to do the calculation. Now Δt_{min} is the minimum time in principle, but if you take your time and pull the state through gracefully, it takes a little longer because it does a lot of backing and hauling. This makes the time ratio in the equation smaller and less entropy is generated.

Because the general formula for energy loss is the entropy times kT, we can write:

$$E = kTq(v_f/v_r),$$

where E is the energy lost per step and the time ratio has been changed to a velocity ratio with v_f the *net* forward velocity of the drifting state and v_r the actual velocity of the state's movement as it sloshes in either direction. From this equation, we see that the energy loss per fundamental step is not $10^{10}kT$ as it's now designed. It's not the $100kT$ that the DNA is able to accomplish. It's not even $1kT$ or the $(\ln 2)kT$ that von Neumann thought it was. It's, in fact, kT times a smaller number. First, the velocity ratio shows that if you pull the state forward gracefully, you can get very little energy loss relative to kT. The result is essentially the same as Bennett's formula, but we have an additional factor, q, that can be thought of as a coasting factor. The coasting factor tells you how fast you can go through the counter without getting scattered back, and it means you have even less energy loss. The designs of Bennett generally operated with diffusion without coasting. He didn't have inertia in his designs, but I think he was fully aware of the fact that if he included inertia he'd have this extra factor. At any rate, the main point is that the quantum mechanical computer has exactly the same formula for energy loss as the classical computer: it approaches zero in the limit of slow, reversible operation.

Various people have raised objections. They say that if you're going to get the state through the counter in a certain amount of time, you've got to put in a certain amount of energy. They base their objection on the uncertainty principle with its restriction on the relationship between the amount of energy in the system versus the time it remains there. As a result they analyze the energy needed in a quantum mechanical system to transmit a certain amount of information in terms of the frequency used, the bandwidth, and so forth. But you put that energy into the original signal, and when it comes out the other side it's still there. It can be recovered and pumped back into the system. What we really need to worry about is not the absolute energy but the energy that's lost, that is, the free energy that produces chaos or irregularities. This concern is just like what somebody said to me about my atomic computer. You've estimated the energy, he said, but you've forgotten how much energy it takes to manufacture the computer. Well fine. Let's say that's the mc^2 of all the atoms. But when I'm done computing after 100 years, I'm going to get that energy back again.

3 Past, Present, and Future of Nuclear Magnetic Resonance

F. Bloch

Like almost every advance in physics, nuclear magnetic resonance (NMR) owes its origin to a chain of preceding developments. An important link in this chain was forged fifty years ago by Otto Stern; although many nuclei had already been understood from the hyperfine structure of spectral lines to possess a characteristic electric and magnetic moment, the magnetic moments of the proton and the deuteron were not known until their magnitude was determined by Stern and I. Estermann from the deflection of a molecular beam of hydrogen in its passage through an inhomogeneous magnetic field. The study of nuclear moments in molecular and atomic beams was subsequently taken up at Columbia by I.I. Rabi and his collaborators. They were able during the following years to gradually extend the power of the method by a series of innovations which culminated in the introduction of nuclear magnetic resonance.

This decisive step forward was based upon a paper of Rabi, published in 1937, where he treated the transition between the Zeeman-levels of a magnetic moment μ in a constant magnetic field H under the influence of a field rotating at right angles to the constant field. The transition probability was shown to reach its maximum under the resonance condition

$$H\mu = s\hbar\omega \tag{1}$$

27

NEW DIRECTIONS IN PHYSICS
The Los Alamos 40th Anniversary Volume
ISBN 0-12-492155-8

where ω is the circular frequency of the rotating field, typically in the range of radio frequencies, and s the spin associated with the magnetic moment. The requirement of a rotating field can be bypassed since the simpler application of an alternating field calls for only a minor and calculable correction. To observe the effect in a molecular beam, the molecules pass from a region of the homogeneous field H into a region with a strongly inhomogeneous field so that the occurrence of transitions in the former is detected by a change of deflection in the latter. With the condition for resonance thus established and given the spin of a nucleus, the magnetic moment is then obtained from Eq. 1 by the strength H and the frequency ω of the constant and of the alternating perpendicular field respectively. The accuracy is limited in principle only by the flight time Δt of the molecule through the transition region, leading to an indeterminacy

$$\Delta\omega \cong 1/\Delta t \qquad (2)$$

of the resonance frequency but still allowing a far more precise determination of nuclear moments than had been possible by the earlier methods.

A second use of magnetic resonance was made shortly before the war in an experiment by L.W. Alvarez and the author to measure the magnetic moment of the neutron. As an essential difference from the application to molecular beams, however, the scattering of slow neutrons in magnetized iron, rather than the deflection in an inhomogeneous field, was observed to ascertain the establishment of resonance conditions.

The preceding account may seem to merely deal with the prehistory of nuclear magnetic resonance since the abbreviation "NMR" is now customarily used in connection with the method introduced after the end of the war by E.M. Purcell and the author. Although the observation of resonance is still essential in some of the most important applications, it is not this feature which characterizes the method and there are in fact other relevant instances where resonance is of secondary significance or even entirely absent. The real distinction from earlier methods has to be seen in the novel principle of detection, based upon purely electromagnetic effects, which allows the investigation of nuclear magnetic phenomena in bulk matter and under many different aspects.

It is indicated, under these circumstances, to describe the principle in terms of the macroscopic polarization resulting from the magnetic moments of the nuclei in a sample. Whereas the vector of polarization at thermal equilibrium in a constant magnetic field is static and parallel to that field, its essential dynamic manifestations appear under conditions chosen to result in deviations from parallelity. The polarization is seen, thereupon, to undergo a precession around the direction of the constant field H with the "Larmor frequency"

$$\omega_L = \gamma H \tag{3}$$

where

$$\gamma = \mu/s\hbar \tag{4}$$

is the "gyromagnetic ratio" of the nucleus. An alternating voltage of the same frequency is thus induced in an external circuit and it is the signal, due to this voltage, combined with the normal methods of amplification, which lends itself to direct observation. To emphasize this truly distinctive feature, the term "nuclear induction" was originally proposed but it is now used only in reference to free precession, observed under nonresonant conditions.

As in the earlier applications, an alternating field, perpendicular to the constant field, can lead to resonance. In regard to the polarization, the resulting action is related to free precession in a manner analogous to that of the forced to the free oscillation of the harmonic oscillator with the resonance condition of Eq. 1 seen to be satisfied when the frequency ω of the alternating field equals the Larmor frequency ω_L of Eq. 3. The angle of the vector of polarization against the constant field reaches here its maximum, indicated by a maximum of the observed signal. In contrast to the situation in a molecular beam, the accuracy is not limited by the flight time since the nuclei remain in the sample under investigation but there are other causes of an error, comparable to the linewidth $\Delta\omega$ of the resonance.

An obvious cause concerns the homogeneity of the constant field; if H varies by an amount ΔH over the extension of the sample, the resonance frequency of a nucleus, depending upon its location, will vary in view of Eq. 1 by the corresponding amount

$$(\Delta\omega)_H \cong \gamma\Delta H \tag{5}$$

with γ from Eq. 4. Another more fundamental cause arises from the random processes which establish the thermal equilibrium of nuclear moments in the constant magnetic field through interaction with their environment. It is necessary for the effect of these processes upon the vector of polarization to distinguish between the approach towards the equilibrium value of the component parallel to the field and the disappearance of an initial perpendicular component. Depending upon the specific properties of the interaction, the two components are found in general to change at a different rate with the inverse characterized by a "longitudinal relaxation time" T_1 for the parallel component and by a separately observable "transverse relaxation time" $T_2 \leqslant T_1$ for the perpendicular component which leads in the observation of resonance to a natural linewidth

$$(\Delta\omega)_n \cong 1/T_2. \tag{6}$$

The basic facts, outlined above, were well understood already in the first demonstrations of NMR, carried out on the protons in paraffin and water. While they merely confirmed with comparable accuracy the value of the magnetic moment, known from magnetic resonance in a molecular beam, not even the order of magnitude of the relaxation time had been surmised until T_1 was found to be about three seconds in water. This result was of particular relevance since there were good reasons in a liquid of low viscosity to expect the equality of T_1 and T_2 and, hence, from Eq. 6 to envisage a natural linewidth of no more than a fraction of a cycle per second. Although the first experiments were carried out in a relatively weak field H, corresponding to a resonance frequency of only a few megacycles per second, such a small linewidth would in principle have allowed the determination of the magnetic moment with an accuracy of about one part in ten millions. The width of the observed resonance, however, was much larger and entirely due to the inhomogeneity of the field according to Eq. 5; it became clear, therefore, that greater homogeneity should be attempted, even without realizing at the time that this would lead to further developments of major significance.

The rather primitive instrumentation of the original experiments was soon replaced by better equipment. Besides improvements in the circuits and electronics, the stronger and more homogeneous fields, obtained by specially designed magnets, were found to greatly enhance the quality of the observed signals. Among the immediate benefits of higher sensitivity and resolution thus achieved, it was possible to determine the magnetic moment of many nuclei either with greater accuracy than had been previously attained or in cases where it had not been known before. Most of the measurements were carried out on samples in which the molecules, containing the nuclei under investigation, were dissolved in a liquid in order to obtain resonances of sufficiently small linewidth. Another early application of NMR concerned the structure of crystals. In contrast to a liquid one observes here broad resonances due to the fact that the effective field H, acting upon a nucleus, varies over a range determined by the different orientations of neighboring nuclear moments. From the magnitude of the broadening it is possible, therefore, to conclude upon the distance between neighbors. The rapid change of relative positions in liquids causes such a type of broadening to be averaged out and it is because of this "motional narrowing" that liquid samples offer distinct advantages when high resolution is of importance.

The usefulness of NMR had already been widely recognized, and it was actively pursued in many laboratories within a few years after the announcement of the first results. It would lead too far, even in bare outline, to describe the great variety of investigations where NMR was fruitfully applied, first in physics but soon also in other branches of science. A few examples will be chosen, instead, to indicate at least some of the lines along which the developments have moved and are still in progress.

Once the magnetic moment of the proton had been known with considerable accuracy, NMR was used conversely as a convenient method for the measurement of magnetic fields. In an interesting version put to practice, the protons in a hydrogenous liquid are first allowed to reach the equilibrium polarization in a relatively strong field at right angles to the earth's field to be measured. Upon quick removal of this field the polarization, still perpendicular to the earth's field, begins its free precession around the direction of the latter and induces an alternating voltage in a coil surrounding the sample. Due to a long relaxation time, the slow exponential decrease of the amplitude allows the timing of several thousand cycles, equivalent to a correspondingly accurate determination of the Larmor frequency ω_L, with the result for the magnitude H of the earth's field obtained from Eq. 3 and the known value of γ.

For the measurement of stronger fields, the observation of resonance at a frequency ω for protons similarly yields the value of H from Eq. 1. Irrespective of this value, however, one obtains the magnetic moment of a nucleus with known spin s in units of that of the proton moment from the ratio of their respective resonance frequencies, observed in the same field. The magnetic moment of the proton has thus served as a standard and has been calibrated for the conversion of the data into conventional units.

In the course of measuring the magnetic moment of different nuclei, the discovery of the "chemical shift" came as a byproduct. It manifests itself in a slight dependence of the resonance frequency on the compound in which a given nucleus is observed and is due to a shielding of the external magnetic field by the electrons of the molecule. Actually, it is only the relatively small contribution to the shielding by the valency electrons that varies with the compound, but the resulting shifts are sufficiently pronounced and characteristic that they have become an important indicator in chemical analysis. A chemical shift was found not only between different compounds but also between different bonds within the same molecule, a fact which has led to the extended use of NMR for the investigation of molecular structures.

The first evidence of this fact appeared upon the introduction of pulse methods for the observation of NMR signals. Instead of the steady state under resonance conditions, one deals here with the transient response of the polarization to an intermittent action of the alternating field. For a pulse of a given short duration the action can be seen to consist in a rotation of the vector of polarization against the direction of the constant field by an angle proportional to the amplitude of the alternating field. In particular, the application of a 90° pulse to a sample in an inhomogeneous field, followed by a delayed 180° pulse, results after the same further delay in the appearance of a signal called a "spin echo". The magnitude of the echo was found for some substances to vary with a change of the delay time in a manner which could be related to internal properties of the molecule. It suggested the possibility of reaching equivalent conclusions more directly from data to be obtained with sufficiently high resolution from the observation of resonances.

The need for high resolution became evident in the studies on ethyl alcohol which provided the basis for all later research directed at the structure of molecules by means of NMR. In fact, it required very great homogeneity of the magnetic field, as indicated by Eq. 5, to reduce the linewidth to about 100 cycles per second, comparable to the chemical shift of three different groups of protons in the molecule. A further reduction, down to the natural linewidth of no more than one cycle per second, was achieved by rapid rotation of the sample which effectively suppresses the broadening by the remaining inhomogeneity. At this point a large number of closely spaced resonances could be resolved which arise from a very slight coupling between the magnetic moment of protons in neighboring groups.

This type of "spin–spin splitting" is found for a great variety of compounds in patterns which are highly characteristic of the molecule and occur in the resonances both of protons and of other nuclei typically present in organic molecules. Combined with the chemical shift, these patterns convey a good deal of essential information such as the spatial arrangement of different radicals within the molecule. For this reason as well as for its non-destructive character and other advantages, the application of high resolution NMR has been extended to almost the entire field of organic chemistry. In a natural further extension, NMR has become increasingly important to biochemistry and biology where it has the special merit to allow investigations *in vivo* without interfering with the function of the organism.

The significance of NMR as a valuable tool for research continues to grow both in scope and in quality. There are particularly two innovations which contribute at present to a considerable improvement in the performance of the equipment. The first is due to the attainment of magnetic fields, both more stable and considerably stronger than those obtained from electro-magnets, by means of the persistent current in a superconducting coil. Since the signal greatly increases with increasing field and correspondingly increasing frequency, this allows observation even in cases where only small amounts of the sample substance are available. In addition, the chemical shift is proportional to the field so that a greater spread of the resonances can thus be achieved.

A second important contribution is made by computers through their extensive use in the processing of data. In a frequent procedure the alternating field operates in a series of widely spaced 90° pulses and the computer is programmed to produce the Fourier transform of the transcient signal which follows each pulse due to the free precession of the polarization. By this process the time-dependence of the signal is converted into signal strength versus frequency so that even the result, derived from a single pulse, is the same as that otherwise obtained by scanning through the entire sequence of resonance frequencies. The signal-to-noise ratio is improved, however, by accumulation and averaging over a number of pulses. Besides furnishing a display of the transformed signal, the computer can be further programmed to extract relevant numerical values such as those of chemical shifts and spin–spin coupling constants, characteristic of an organic molecule.

The applications of NMR in physics, chemistry, and biology are newly expanded to provide diagnostic information in the field of medicine. Through the introduction of sufficiently large magnets it is now possible, in particular, to investigate specific metabolic processes in the human body by means of high resolution NMR. The chemical shift between the phosphorus metabolites, observed in the so-called "topical magnetic resonance" of the isotope ^{31}P, thus allows to determine their relative concentration and hence to obtain a non-intrusive test for the muscle function in the arm or the leg.

One of the most promising clinical applications of NMR, however, is due to the remarkable progress in a recent development, named "NMR imaging" or "zeugmatography" from the Greek "zeugma", meaning that which joins together. What is here being joined is the frequency of the signal and the location of the nuclei from which it originates. The principle is a direct consequence of Eq. 3 with the difference that H now refers to an inhomogeneous field which has a known dependence of its magnitude on the space-variables. Considering, in particular, protons, all those observed to have a given Larmor frequency ω_L are therefore known from their value of γ to be located on a surface

$$H(x, y, z) = \omega_L/\gamma = \text{constant.} \tag{7}$$

There are several schemes, based upon this principle, to arrive at the spatial distribution of protons and thereby at an image representing the shape and location of tissues within which they are contained. In one of the procedures the method, described above, is applied to obtain the frequency distribution from the Fourier transform of the transient signal following a 90° pulse. For a constant gradient of H with the surfaces of given frequency thus being parallel planes, this constitutes a one-dimensional projection of NMR responses along the direction of the field gradient. By rotating the direction, data of different projections can then be processed to give a two-dimensional image in analogy to the processing used in x-ray tomography (CT).

While NMR imaging can at present resolve details only to within a few millimeters and is in this respect inferior to CT scanning, it avoids the damage to cells caused by x-rays. As another distinguishing feature of considerable importance, it allows us to discriminate between different tissues such as the gray and the white matter in the brain or to recognize a tumor through the difference not only in the density but also in the relaxation time T_1 as well as T_2 of their protons. It was found furthermore that malignant tissues generally have a longer relaxation time than normal tissue; although there is some overlap, this difference may prove to be of significance in the detection of cancer.

The fact that NMR has long since been firmly established in many fields of endeavor leaves little doubt that its use, aided by the availability of specially designed instrumentation, will extend into the foreseeable future. Nor are there reasons to believe that all potentialities of the method have been exhausted and that major new advances are no longer possible. Given the wide interest in the application to medicine, for example, much progress can be expected as more experience will be gained to correlate clinical evidence with the acquisition of NMR data.

A good deal of the general prospects for further developments depends upon the extent to which one will succeed in the attainment of stronger NMR signals. This is of special importance for elements other than hydrogen since both the abundance of suitable isotopes and the magnitudes of their magnetic moments compare unfavorably with those of the proton. Besides operating at the highest magnetic fields and the lowest temperatures that are feasible, it is possible in some instances to increase the signal by applying simultaneously with the alternating field for NMR a field at a higher frequency chosen to cause transitions between certain states which are coupled to the Zeeman-states of the observed isotope. The effect is the more pronounced the higher the frequency and it can lead to an increment of the signal by several orders of magnitude where transitions between electronic states are involved. In order to fully exploit such opportunities, the combination of NMR not only with microwave but also with laser techniques may thus become an important auxiliary device.

It is difficult to anticipate the directions in which NMR is going to develop and there is no assurance for the previous expansion to continue at the same rate. The prospects are not to be underestimated, however, if experience offers any guidance about the future of NMR; all expectations at the time of its modest beginnings were so far surpassed during the following decades that one should be prepared for more surprises to come.

4 Experimental Evidence That an Asteroid Impact Led to the Extinction of Many Species 65 Million Years Ago[1]

Luis W. Alvarez

I used to be able to say just about everything I know about this subject in an hour. I could develop it in historical order in a standard-length lecture, but things have moved so rapidly in the last two years that it is quite impossible to follow that scheme anymore. Therefore, I am going to have to concentrate on the present state of our theory that an asteroid hit the earth 65 million years ago and wiped out large numbers of species, both on the land and in the ocean.

I think the first two points—that the asteroid hit, and that the impact triggered the extinction of much of the life in the sea—are no longer debatable points. Nearly everybody now believes them. But there are always some dissenters. I understand that there is even one famous American geologist who does not yet believe in plate tectonics and continental drift. We now have a very high percentage of people in the relevant fields who accept these two points. Of course, science is not decided by a vote, but it has been interesting to watch the consensus develop.

[1] Presented at the annual meeting of the National Academy of Sciences, Apr. 28, 1982, Washington, DC.

NEW DIRECTIONS IN PHYSICS
The Los Alamos 40th Anniversary Volume
ISBN 0-12-492155-8

Reprinted from *Proc. Nat. Acad. Sci.*
80:627–642, Jan. 1983.

The third point, that the impact of the asteroid had something to do with the extinction of the dinosaurs and of the land flora, is still very much open to debate, although I believe that it very definitely did. But I will tell you about some of our friendly critics who do not. I will concentrate on a series of events that has led to a great strengthening of the theory. In physics, theories are declared to be strong theories if they explain a lot of previously unexplained observations and, even more importantly, if they make lots of predictions that are verified and if they meet all the tough scientific challenges that are advanced to disprove them. From that process, they emerge stronger than before.

I am going to present a number of predictions that our theory has made; almost without exception they have been verified. And I will tell you of several serious questions and doubts that have been raised concerning the validity of the theory. People have telephoned with facts and figures to throw the theory into disarray, and written articles with the same intent, but in every case the theory has withstood these challenges. I will therefore concentrate on those things that show the theory to be a strong one, but I will not neglect a few "loose ends".

Instead of using the historical approach, which has been my custom up till now, I am going to start by following the cub reporter's checklist that he learned in journalism school. Every story should contain Who, What, When, Where, and Why.

First of all, Who? The original "Who" were the Berkeley group (Fig. 4.1). Let me introduce my colleagues, shown in alphabetical order, in our first major publication.[2] The second one is my son, Walter, who is a professor of geology at Berkeley. Frank Asaro and Helen Michel are nuclear chemists at the Lawrence Berkeley Laboratory. All of us have been involved in every aspect of the problem, since the earliest days. I have even been out looking at some rocks in Italy—a new experience for me. Helen Michel has collected rock samples in Montana, where there are dinosaur fossils. Her husband tripped over a previously undiscovered Triceratops (horned dinosaur) skull on one occasion. We have not been a group of people each working in his own little compartment, but rather we have all thought deeply about all phases of the subject.

When we sent the paper [1] to *Science*, Philip Ableson, its editor, had two comments. In the first place, it was too long. He could not publish it unless we cut it in half. It still turned out to be pretty long. Second, Phil said, "I have published quite a few papers on the cause of Cretaceous-Tertiary extinction in the last few years, so at least $n - 1$ of them have

---

[2] Alvarez, L.W., Alvarez, W., Asaro, F. & Michel, H.V. (1979) Lawrence Berkeley Laboratory Report 9666.

EXTRATERRESTRIAL CAUSE FOR THE CRETACEOUS-TERTIARY
EXTINCTION: EXPERIMENT AND THEORY

Luis W Alvarez*[+]

Walter Alvarez**

Frank Asaro*

Helen V. Michel*

University of California
Berkeley, California 94720

*Lawrence Berkeley Laboratory

**Department of Geology and Geophysics

[+]Space Sciences Laboratory

XBL 829-9695

Figure 4.1. Title page of Lawrence
Berkeley Laboratory Report 9666.

to be wrong." But, in spite of that, he published ours, and we are most
appreciative.

Since we first presented our results at three geological meetings
[2–4] starting in early 1979, about 12 other groups have entered the field.
The latest one to be heard from is a Russian group.

Now, the "What" category. We have very strong evidence that an
asteroid (Fig. 4.2) hit the earth 65 million years ago at a velocity in the
range of 25 kilometers per second. You may wonder how we got this
picture of an asteroid that hit the earth 65 million years ago. Actually,
this is a picture of Phobos, the larger of the two moons of Mars. It was
taken by the Mars Orbiter, and I was surprised to see that it was pocked
with craters. I had always imagined "our asteroid" as being a nice,
smooth, round thing that ran into the earth, but of course it must have
been bumped into by many, many smaller asteroids and meteorites, so
this is what it undoubtedly looked like. Phobos is actually twice the size
of "our" 10-kilometer-diameter asteroid, but otherwise it looks exactly
the same. From the color of Phobos NASA found that it is probably a
carbonaceous chondrite, and we have very strong evidence that the
asteroid that hit the earth was also of carbonaceous chondritic composition.

When the asteroid hit, it threw up a great cloud of dust that quickly
encircled the globe. It is now seen worldwide, typically as a clay layer
a few centimeters thick in which we see a relatively high concentration
of the element iridium. This element is abundant in meteorites, and
presumably in asteroids, but is very rare on earth. The evidence that

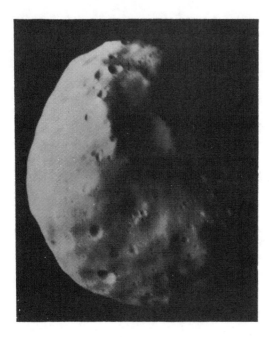

Figure 4.2. Phobos, a satellite of Mars. (Photo courtesy of National Aeronautics and Space Administration.)

we have is largely from chemical analyses of the material in this clay layer. In fact, meteoritic iridium content is more than that of crustal material by nearly a factor of 10^4. If something does hit the earth from outside, you can detect it because of this great enhancement. Iridium is depleted in the earth's crust, relative to normal solar system material, because when the earth heated up and the molten iron sank to form the core it "scrubbed out" the platinum group elements in an alloying process and took them "downstairs". (We now use the trick of heating our rock samples with molten iron, to concentrate the iridium, and thereby gain greatly in signal-to-noise ratio.)

We come to "When" on the checklist. There are two time scales for the "When". The first one is the geological time scale (Table 4.1), which I now know the way I know the table of fundamental particles. Note that the 570-million-year time span from the beginning of the Cambrian up to now is called "Phanerozoic time"—that is, when there are easily observed fossils in the rocks. Phanerozoic time is divided into three eras: the Paleozoic, or old animals; the Mesozoic, or middle animals; and the Cenozoic, or recent animals. The fact that geologists characterize their rocks by the fossils that are in them shows us the close interrelationship between geology and paleontology.

I am going to concentrate most of my attention on what could be

Table 4.1. Geological time chart of the Phanerozoic eon

Era	Periods or system*	Epochs or series†	Time since beginning, yr
Cenozoic	Quaternary	Recent	
		Pleistocene	1,000,000
	Tertiary	Pliocene	12,000,000
		Miocene	30,000,000
		Oligocene	40,000,000
		Eocene and Paleocene	65,000,000
Mesozoic	Cretaceous		120,000,000
	Jurassic		155,000,000
	Triassic		225,000,000
Paleozoic	Permian		250,000,000
	Carboniferous		300,000,000
	(Pennsylvanian; Mississippian)		
	Devonian		350,000,000
	Silurian		390,000,000
	Ordovician		480,000,000
	Cambrian		570,000,000

* Period of time or system of rock.
† Epochs of time or series of rock.

called the Mesozoic-Cenozoic boundary, but everyone calls it the "Cretaceous-Tertiary (C-T) boundary". It is 65 million years old. I will also talk briefly about the Permian-Triassic (P-T) boundary. That is when there was another major extinction. I should say that there have been five major extinctions in Phanerozoic time [5]. I will also say something about the boundary between the Eocene and the Oligocene which is at about 34 million years ago and was accompanied by a less-severe extinction event.

Raup and Sepkowski [6] recently published a definitive article on the five major extinctions, from which I have used a plot of the number of extinctions at the family level, per million years, against time (Fig. 4.3). Such a graph makes me feel right at home because for a good many years I was called a "bump hunter"—a particle physicist who looks for "resonances" or peaks that stick out above a distribution of background points. Figure 4.3 shows that there *is* a substantial background of extinctions; individual families are going extinct all the time, for natural reasons quite unconnected with the events that have triggered the five "major extinctions". And those who criticize our asteroid theory of the C-T extinctions have known about this background for much longer than I have. But I think that on many occasions they have, as we would say in physics, confused some background events with events that really

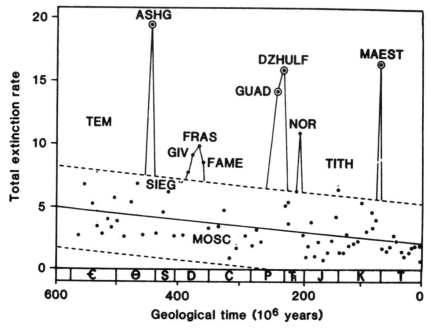

Figure 4.3. Total extinction rate (extinctions per million years) through time for families of marine invertebrates and vertebrates. The plot shows five mass extinctions: late in the Ordovician (ASHG), Devonian (GIV-FRAS-FAME), Permian (GUAD-DZHULF), Triassic (NOR), and Cretaceous (MAEST) periods. The late Devonian extinction event is noticeable but not statistically significant. (From Raup and Sepkowski [6], by permission of the American Association for the Advancement of Science, copyright 1982.)

belong to the peak. I mention this because I believe that such a confusion has contributed to the present controversy concerning the validity of the asteroid hypothesis. When we point to a number of species that went extinct precisely at the iridium layer, our critics commonly discount those extinctions by pointing to other species that were obviously "on the way out" just before the asteroid hit. That is what I call "confusing the background with the peak events," and, if I did not direct attention to this graph, you might find those arguments against our theory more persuasive than the evidence warrants.

The second time scale is the present time scale and is concerned with the discovery of iridium enhancements in the geological record and with their interpretation in terms of an asteroid impact. We started our search five years ago. We saw our first iridium "spike" four years ago. We were looking for iridium but, it turns out, for the wrong reason. The first time we saw the iridium enhancement we did not have a sufficiently complete set of rock samples, so Walter went back to Gubbio, Italy, and collected the set whose analysis makes up the points shown in Figure 4.4. We plotted that curve three years ago and showed it at a number of geological meetings. This is an unusual diagram, with time plotted upward, in a linear mode in the middle section and in a logarithmic mode in the top and bottom sections. The iridium concentration, which has been fairly constant for 350 meters below the C-T boundary, increases sharply—by a factor of about 30—in the 1-cm clay layer and then decreases as one goes into the earliest Tertiary limestones. For the rest of the 50 m above the boundary, the iridium concentration is at the background level seen in the late Cretaceous limestones.

This is the very large signal that we explained as being due to the impact of an extraterrestrial object. If I were following the historical approach, I would give you our original justification for that conclusion. But instead I will later give you more recent data that show beyond any question that the clay layer contains "undifferentiated" solar system material, with a composition that matches that of carbonaceous chondrites with surprising accuracy. Our first thought was that the material came from a supernova because some paleontologists believed at that time that a nearby supernova was responsible for triggering the C-T extinction. But we soon found that the clay was too similar to solar system material to be from a supernova. I sent a letter to Malvin Ruderman, a physicist friend of mine and one of the key exponents of the supernova theory, explaining why we could no longer accept his theory. He wrote back a very short letter saying, "Dear Luie: You are right, and we were wrong. Congratulations. Sincerely, Mal". That is something that made me very proud to be a physicist, because a physicist can react instantaneously when you give him some evidence that destroys a theory that he previously had believed. But that is not true in all branches of science, as I am finding out.

Three years ago we had this graph and this theory. We wrote it up, and it was published in *Science* [1]. Now, a little more on "When". Since our original work, there have been three conferences on the subject, because it is such a rapidly evolving field. The first conference [7] was held about one year ago in Ottawa under the sponsorship of the National Museums of Canada; about 25 people were there, people who study meteorites, impact craters, geology, paleontology, and quite a range of subjects, and we had a very good three-day meeting.

Last fall there was a four-day meeting at Snowbird, Utah. It was sponsored by the Lunar and Planetary Institute and by the National Academy of Sciences. One hundred and ten people attended that meeting, which lasted four days. They came from more fields than you can imagine,

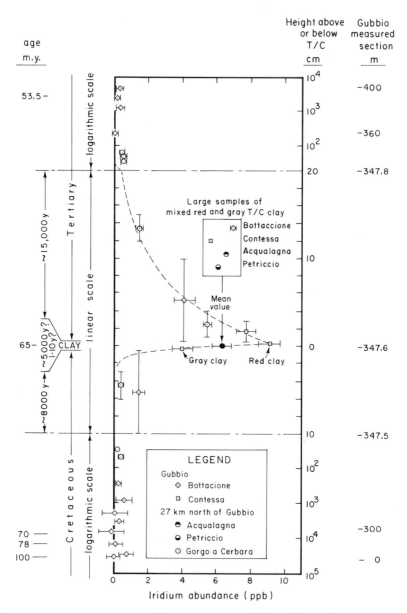

Figure 4.4. Iridium abundance per unit weight of acid-insoluble (2 M HNO₃) residues from Italian limestones near the C-T boundary. Error bars on abundances are SD for counting radioactivity. Error bars on stratigraphic positions indicate the stratigraphic thickness of the samples. The dashed line is an "eyeball exponential fit" to the data.

including atmospheric modeling, impact dynamics, chemistry, physics, asteroids, and, of course, geology and paleontology. We had a very good exchange of views, and almost everyone in this new field had a chance to meet "all the players". More recently, there was a day-long seminar [8] on this subject at the 1982 meeting of the American Association for the Advancement of Science.

Now, the "Where". The iridium enhancement was first seen near a little town called Gubbio, which is in the north central part of Italy. It is directly north of Rome and directly east of Siena in the Apennines. The rocks there were laid down as limestone on the bottom of the ocean from 185–30 million years ago, and then a few million years ago they were raised up in the mountain-building process. They were then eroded by running water, and fortunately engineers built roads up through the canyons so that someone like me, an armchair geologist, could get there in comfort. I found that I could get out of the car, wield a geologist's hammer to break a new surface of the rock, and look at the little creatures that lived there and see how they changed with time.

It is really dramatic to observe the little things called foraminifers (order *Foraminifera*) which are shelled creatures about 1 mm in diameter (Fig. 4.5b). You can see them with a hand lens literally by the thousands, right up to the boundary, at apparently constant intensity, and then, without warning, they are gone, right at the clay layer. It was really a catastrophe. They were suddenly wiped out. The only foraminifers that escaped extinction were the tiny species *Globigerina eugubina* that can be seen in the thin section, above the boundary line in the same figure.

Figure 4.6 shows what the rocks look like. This layer was deposited 65 million years ago, and it is seen many places worldwide. We took it upon ourselves to analyze the layer by neutron activation analysis, looking particularly for iridium. You have already seen the iridium enhancement, which surprised us so greatly when we first saw it in 1978.

The limestone in this region is about 95% calcium carbonate and about 5% clay. The calcium carbonate comes from the shells of the little animals that live in the ocean and fall to the bottom when they die. The clay is washed down from the continents and carried out to sea by river currents. The two components fall to the ocean floor, where they are compacted to form the limestone.

It was generally assumed, before we did our work, that the clay in the layer was of the same origin as the clay in the limestone, but that turns out not to be the case. After we had seen the iridium in the layer and concluded that it came from an asteroidal impact, we made our first prediction—that the gross chemical composition of the clay layer would

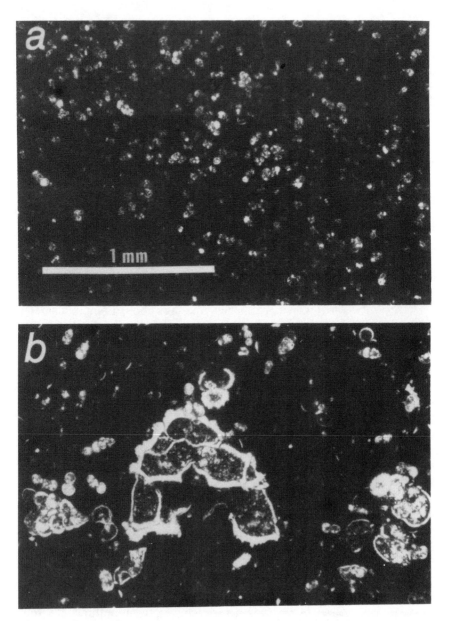

Figure 4.5. Photomicrographs from the Bottaccione Section at Gubbio. (*a*) Basal bed of the Tertiary, showing *Globigerina eugubina*. (*b*) Top bed of the Cretaceous, in which the largest foraminifer is *Globotruncana contusa*.

Figure 4.6. (*Upper*) L.W.A. (left) and W.A. pointing to the C-T boundary in the Bottaccione Gorge near Gubbio, Italy. (*Lower*) Close-up of the C-T boundary, with a coin (similar to a U.S. quarter) indicating the size of the boundary.

be substantially different from that of the clay in the Tertiary and Cretaceous limestones above and below the layer and that these latter two clays would be essentially identical. We published measurements in our *Science* paper that showed that this first prediction was verified.

Our second prediction was that the iridium enhancement would be seen worldwide. At that time we had only seen it in one place in Italy, in a valley near Gubbio. We knew that the extinctions were worldwide, and our prediction that the iridium would be seen worldwide turned out to be true as well. Before we published our 1979 paper, we had samples from Denmark, which Walter collected, and also some from New Zealand that Dale Russell was kind enough to give us. Both of those showed a nice iridium enhancement. Both enhancements were bigger than the one we saw in Gubbio. In fact, as shown in Figure 4.7, we first discovered the iridium in nearly the hardest place to find it, where the iridium concentration was quite small compared to most places. The number indicated for each site is the measured iridium, in nanograms per square centimeter, at that location. This is of course the area under the curve of the type in Figure 4.4, times the density of the rock.

At the present time, there are more than 36 locations where the iridium has been found. With one exception the iridium has been found every place that has been thoroughly looked at by our laboratory. Whenever

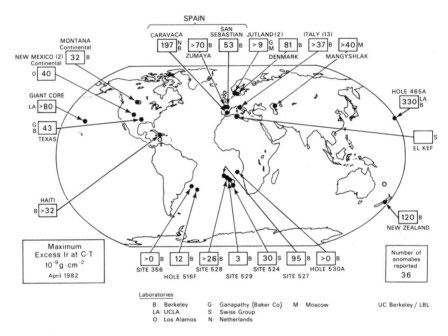

Figure 4.7. Map of the world with locations of iridium anomalies. Laboratories: B, Berkeley; LA, University of California at Los Angeles; O, Los Alamos; G, Ganapathy (Baker, Co.); S, Swiss group; N, Netherlands; M, Moscow.

a paleontologist says, "This is the C-T boundary", one of the groups now looking for iridium collects some rock samples and finds the iridium enhancement by using neutron activation analysis. The one place where this is not true is in Montana. We have two sites in Montana where there are abundant dinosaur fossils. But it is not so easy to pick out the C-T boundary, and there is no obvious clay layer. (The clay layer is seen in nearly all the marine deposits.) In one of these Montana sites, we have iridium, but we have not found it at the other site, even after two summers of sample collecting. Thus, it is almost correct to say that iridium has been found at every identified C-T site that anyone has looked at. In all of the pelagic or ocean-based sites, the iridium was laid down on the ocean floor 65 million years ago, and it has been found, in all our studies, within 10 cm and often within 1–2 cm of the place where the paleontologists said we should look.

I think it is interesting that, after seeing the iridium at one site in Italy, we predicted that it would be seen worldwide at the C-T boundary, and Figure 4.7 shows that that prediction has been fulfilled. You will see that there are sites in both oceans for which deep sea drilling cores have been made available to us and to other groups. The largest amount of iridium (in the north central Pacific) is 330 ng/cm^2. As a physicist, I had expected that, when we got a map like this, we would be able to draw lines connecting places with equal iridium values and then we would be able to mark the center, as on a contour map, and say, "This is where the asteroid hit." But that is not the way things work in the much more complicated world of geology.

Now, we come to "Why", the last item on the checklist. Why did we study this problem in the first place? I do not really have to explain that to an audience of this kind. If I did, I would probably use George Mallory's famous response as to why he tried to climb Mount Everest, "Because it is there." But if I wanted to get more serious, I would say that, a few years ago, the four of us suddenly realized that we combined in one group a wide range of scientific capabilities, and that we could use these to shed some light on what was really one of the greatest mysteries in science—the sudden extinction of the dinosaurs. How many species or genera went out 65 million years ago? I get a different set of numbers from every paleontologist I talk to, but everyone agrees that it was simply a terrible catastrophe. Most of the life on the earth was killed off; about half of all the genera disappeared completely, never to be seen again.

At this point some comment about the disappearance of the dinosaurs is in order. They were reptiles, and the land reptiles went out in a really catastrophic way. In all, there were several orders of reptiles that disappeared completely, including giant marine reptiles. Normally, one talks about an extinction at the species level. The passenger pigeon disappeared in the last century. The condors are probably going out soon. Each is a species extinction. Above that we have a genus or many genera; above that comes the family; above that is the order and, for fauna, the only higher taxa are class and phylum. Thus, an extinction that suddenly wiped out several orders was a spectacular catastrophe—not to be attributed to some ordinary environmental change, as some of my friends believe. Dinosaurs were some of the biggest animals that ever lived on the land, *Tyrannosaurus rex* for example. There were large reptiles in the seas— the plesiosaurs. There were large reptiles that were flying around in the air—the pterosaurs. All disappeared suddenly, never to be seen again. I simply do not understand why some paleontologists—who are really the people that told us all about the extinctions and without whose efforts we would never have seen any dinosaurs in museums—now seem to deny that there ever was a catastrophic extinction. When we come along and say, "Here is how we think the extinction took place", some of them say, "What extinction? We don't think there was any sudden extinction at all. The dinosaurs just died away for reasons unconnected with your asteroid." My biggest surprise was that many paleontologists (including some very good friends) did not accept our ideas. This is not true of all paleontologists; some have clasped us to their bosoms and think we have a great idea.

To this overview of the situation I should just add one point. Dinosaurs did last for nearly 140 million years from the early Mesozoic, which is sometimes called the age of reptiles, and we believe that had it not been for the asteroid impact, they would still be the dominant creatures on the earth. We would not be sitting here. At least we would not look as we do; it has been suggested that we would have distinctly reptilian features.

Now I must add a few odd facts that do not fit into the checklist that I started out with. One is that "earth orbit crossing asteroids" are studied by two groups of people. One group looks at them as astronomical objects by using Schmidt cameras on whose photographic plates the asteroids appear as streaks moving relative to the background stars. The other group studies craters, either on the moon or on the earth. There is some overlap in these two populations. For example, Eugene Shoemaker is an expert in both of these fields.

All of these people agree that there is a power law relationship between the mean time to collision of an asteroid of a given size and its diameter:

the mean time to collision is roughly proportional to the square of the diameter of the object. These two groups of people also agree on the absolute numbers. What they say is that an object 10 kilometers in diameter should hit the earth every 100 million years, on the average. If you drop the size by a factor of 10, to 1 kilometer, then you drop the mean time to collision by a factor of 100, to 1 million years. If you go down to 100-meter objects, these hit the earth about every 10,000 years. That power law goes over an enormous range of sizes. It has been verified on the moon, where you can see very small craters. On the earth, the little craters have been eroded away or the objects burned up in the atmosphere, so you can only see the evidence of the big ones.

There have been five major extinctions in the last 570 million years, and our third prediction was that all of these would turn out to be caused by the same mechanism, an asteroid collision. That is one prediction that has not turned out to be true, but it does have an element of truth. We have only looked at one other of the five major extinctions, the P-T. It is hard to sample, because the best sites are in China. Frank Press, working through our National Academy and the Chinese Academy, helped us get one of the two sets of samples of P-T rocks that we have analyzed. There is a clay layer between the limestone-like rocks at the P-T boundary. We felt sure that there would be lots of iridium there. But there is not any that we can find.

However, we are very intrigued by the existence of that layer, whose basic chemistry is quite different from that of the rocks above and below it. The fact that it exists was not widely known until quite recently; Walter learned about it less than three years ago. Our present best guess is that it is of volcanic origin, but it might be consistent with the idea that the layer was laid down by a cometary impact. Comets can go much faster than asteroids and, in fact, can have 50 times the specific energy. A comet could throw the same amount of dust into the atmosphere and do the same damage, while bringing in only 1% as much iridium. That factor of 100 comes from the square of the increased impact speed times perhaps a factor of 2 because a comet is typically half composed of ice. That is simply one possible working hypothesis. There is no proof for it. But if it does turn out to be true, then we will know that the C-T extinction was due to an asteroid, and not a comet, as some of our friends are calling it. At this point, I think the distinction is of no importance; the important conclusion is that a large chunk of undifferentiated solar system material hit the earth 65 million years ago and triggered a major extinction.

Although our prediction was not confirmed in the P-T case, it did lead to another case in which there is a coincidence between an iridium layer and an extinction, although not one of the five major ones. Some people say, "I'll bet there are lots of iridium layers all over the place, so there is no reason to say that the oceanic and terrestrial iridium layers are synchronous." But in my view, that is an exercise in grasping for straws because it turns out that there are very few iridium layers. No one has yet made a systematic search through all of geological time, but two groups have systematically searched a total of 23 million years of sediments and found not a single iridium enhancement in this randomly selected 4% of Phanerozoic time. One group, led by Frank Kyte and John Wasson of the University of California at Los Angeles, has searched through the lowest 15 million years of the Tertiary limestones; and the other group, led by Carl Orth of Los Alamos, has searched through 8 million years of the late Devonian.

We found a very definite iridium enhancement in the Caribbean Sea, at the Eocene-Oligocene boundary [9],[3] 35 million years ago, and it was independently found by R. Ganapathy [10][4] of the Baker Chemical Company. Both of our groups looked there because that boundary coincided with a known layer of microtektites and with a lesser extinction event. That was very exciting to us because, shortly before we did this work, Billy Glass, a leading expert on microtektites, and his collaborators, had shown that these microtektites—part of the "North American strewn tektite field"—extended more than halfway around the world [11]. And here again, "everybody" (all but one person) believes that tektites are due to the impact of large meteorites (or small asteroids) on the surface of the earth. Also, Billy Glass points out that, at the tektite "horizon", there was an extinction of several species of Radiolaria, much like the foraminifers I talked about earlier, but their chemistry is siliceous rather than calcareous. [Note: In collaboration with Billy Glass, we have recently found three new and quite substantial iridium enhancements at the Eocene-Oligocene tektite horizon in deep sea drilling cores from the Gulf of Mexico, the central Pacific, and the Indian Ocean.]

We have several different bits of evidence that tie impacts to extinctions. We have the iridium layer at the tektite layer and we have the extinction of the radiolarians at that same time. Although we did not find any iridium at the P-T boundary, we did find another iridium enhancement coincident with an extinction, and at the present time there are only two known

[3] Asaro, F., Alvarez, L.W., Alvarez, W. & Michel, H.V., Conference on Large Body Impacts and Terrestrial Evolution: Geological, Climatological, and Biological Implications, Snowbird, Utah, Oct. 19–22, 1981, p. 2 (abstr.).

[4] Ganapathy, R., Conference on Large Body Impacts and Terrestrial Evolution: Geological, Climatological, and Biological Implications, Snowbird, Utah, Oct. 19–22, 1981.

stratigraphic levels where there is a sudden excess of iridium that is seen in more than one location. In regard to the connection between extinctions and impacts, the theory seems to be holding up very well on that score, and the third prediction can be considered to have been partially confirmed. Asteroid impacts have produced more than one extinction but not all five of the major ones.

Prediction number four is that there should be an iridium enhancement at the C-T boundary, on the continents as well as on the sea floor. A lot of people were saying two years ago that the reason we found iridium in the sea floor deposits was that some change in ocean chemistry, 65 million years ago, precipitated out the iridium that was dissolved in the ocean. We had given two arguments in our paper as to why that was not so, but we could not prove it conclusively. We asked one of the national funding agencies for money to search for iridium in Montana, alongside the dinosaurs, and one of the peer reviews that came back said, in effect, "These guys would be wasting their time and your money if they did this job, because the iridium came out of the ocean and therefore won't be seen in continental sites." Fortunately we were able to do it anyway; we went up to Montana and looked for iridium.

But before we got our first iridium there, Carl Orth from Los Alamos and his colleagues discovered that there was iridium at a continental site in New Mexico [12]. I think this was a very important discovery, and I want to show you Carl Orth's curves.

They drilled a hole in the Raton Basin, in New Mexico, and subjected the rocks to neutron activation analysis with a higher sensitivity than anyone else has attained. In Figure 4.8, the scale of iridium abundance is logarithmic. The iridium suddenly went up by a factor of 300 precisely where the paleontologists told them to look. That was a very exciting thing because it showed that the iridium did not come out of the ocean. It was deposited on the continents, as well as on the ocean floor, as called for by our prediction number four. Thus, the Los Alamos discovery added great strength to our theory, as far as some of its critics were concerned. We were not surprised because we thought the arguments we had given against an oceanic source of the iridium were quite valid.

Just a few days before I saw Carl Orth's preprint, which he kindly sent me, I had read a paper [13] by Leo Hickey, who is a paleobotanist in Washington and who has been one of our most vocal critics. He is also a very good friend. Walter went to graduate school with him and they have been close personal friends ever since. His paper in *Nature* was entitled "Land Plant Evidence Compatible with Gradual, Not Catastrophic, Change at the End of the Cretaceous". He wrote this paper after seeing all the evidence that we presented, and I could not find

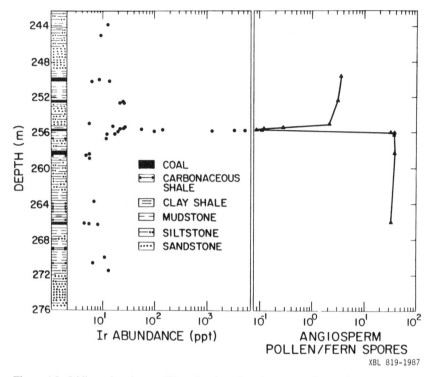

Figure 4.8. Iridium abundances (●) and ratios of angiosperm pollen to fern spores (▲) as a function of core depth and lithology. The surface Ir density is ≈40 ng/cm². (Reprinted from Los Alamos report LA-UR-81-2579.)

anything in it that made me feel that he was ignoring our evidence; he was just looking at a different data base and coming to a different conclusion.

His abstract ends with this sentence: "However, I report here that the geographically uneven and generally moderate levels of extinction and diversity change in the land flora, together with the nonsynchroneity of the plant and dinosaur extinction, contradict hypotheses that a catastrophe caused terrestrial extinctions." His considered opinion after studying all the evidence and looking at what he saw in the plant record convinced him that we were wrong. He says quite clearly that there was no effect of a catastrophe on the plants. And I had no evidence that directly contradicted his conclusions.

You can imagine my excitement when I saw Carl Orth's data as plotted in the right-hand side of Figure 4.8, showing the number of pollen grains per cm³ plotted against stratigraphic height, and normalized to the fern spore count. The interesting thing is that the pollen count drops by a factor of 300, in precise coincidence with the iridium enhancement. I must say that this looks to me like a catastrophe. In fact, several pollen types disappeared from the record at this point. Figure 4.9 shows that the resolution of the pollen fall-off is undoubtedly limited by the sample thickness, 2 cm. The drop-off of a factor of 300 occurs from one rock sample to the next, and my guess is that the discontinuity is even more precipitous than this graph shows. This is by far the sharpest resolution that paleobotanists have ever seen, as far as I can learn, and it is not surprising that it has been missed in the past, just because of its sharpness. (And it confirms prediction number five in our paper—there would be an extinction of plants in coincidence with the iridium layer, on the land.)

To show that the missing of a "sharp spike" is not peculiar to paleobotany, let me remind the physicists in the audience how the psi meson was discovered at Stanford several years ago. The SLAC-SPEAR electron-positron colliding ring had been operating for some time, without anything "very interesting" being found. It was exploration of new territory,

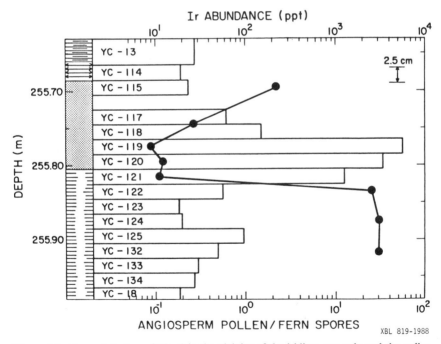

Figure 4.9. Expanded view of Fig. 8 in the vicinity of the iridium anomaly and the pollen break. The Ir abundances are given by the histogram; the angiosperm pollen/fern spore ratios are shown by the solid circles. The lithologic symbols are the same as in Fig. 8 except for coal, which is shown by stipple. (Reprinted from Los Alamos report LA-UR-81-2579.)

and the physicists were looking for enhancements in the counting rate ("bump hunting") by stopping every 100 MeV—equivalent in paleontology to taking a sample every meter. They were unhappily coming to the rather firm conclusion that there were no new "resonances" in this energy region; such resonances were expected to be more than 100 MeV wide. But as a result of some excellent detective work, with attention paid to the slimmest of clues, the SLAC-LBL group looked between a pair of 100-MeV "milestones" and discovered the extremely narrow psi resonance that sent the counting rate up by more than a factor of 100, within the space of 1 MeV and within an observing time interval of two hours. The important point I want to make is that, after those two hours of excitement at Stanford, no one ever said again that there was nothing interesting going on in that wide energy range. The psi "bump" from then on was a part of the lore of physics.

Hickey has behaved quite differently with respect to the "narrow spike" discovered by Carl Orth; he ignored it. At the annual AAAS meeting [8], some months after the Snowbird conference, he repeated the conclusions of his *Nature* paper, that the pollen spectra showed no evidence for a catastrophe, and said, "Every pollen spectrum that has come in since this chart was plotted tends to corroborate these data." However, the narrow "glitch" in the pollen spectrum (Figs. 4.8 and 4.9) contradicts the idea that the plants did not notice the asteroid impact. (My own guess is that, before long, this graph will be reproduced in every textbook on geology and paleontology.)

I consider Orth's important paper to be a confirmation of three separate predictions or deductions we made in our *Science* paper. Prediction number four was that the iridium would be found on the continents as well as on the ocean floor. Prediction number five was that the plants would suffer simultaneous extinctions, jut as the animal life had. And prediction number six was that the iridium did not come from a supernova.

Hickey asserted [8] the plant and dinosaur extinctions were "nonsynchronous" but I think I will soon convince you that he was wrong in that.

Science has published, in the last year, three separate reports on the state of the asteroid theory. They were all written by Richard Kerr. The first one [14], entitled "Asteroid Theory of Extinction Strengthened", reported interviews with people who thought that we were wrong for a number of reasons. I thought the strangest reason was that we found *too much* iridium in the Danish clay layer. Several experts on cratering were quoted as saying that we should not have found nearly that much iridium because, when the asteroid hits, the material going up into the stratosphere should be not only that of the asteroid but also crust material equivalent to 1000 to 10,000 times the mass of the asteroid. Richard Grieve of the Department of Energy, Mines and Resources of Canada was quoted as favoring the figure 1000; Tom Ahrens of California Institute of Technology was said to prefer 10,000. We had used a dilution of 1:60 in our paper, a factor we had gotten from Richard Grieve by telephone a few months earlier. We were, therefore, surprised by Kerr's *Science* report. It turned out later that both of these gentlemen's remarks had been misinterpreted. They both had said that the material close to the crater would be diluted by these very large factors. This ties in well with what we know about Meteor Crater in Arizona—there is very little meteoroid material close to the crater. But both men agreed that the material that was sent up high and spread worldwide would be diluted 1:20 to 1:100, in line with what we observed. This was a major challenge which the theory met and, in the process, was strengthened by. Everybody now agrees that the iridium concentrations we find are consistent with the asteroid impact hypothesis.

Another report by Kerr in *Science* [15], entitled "Impact Looks Real, the Catastrophe Smaller", came after the Snowbird meeting in November 1981 and indicated that a consensus had formed in favor of the asteroid theory. There we had come up against a really serious challenge, involving good science, in which the new numbers were in serious disagreement with the corresponding ones we had used in our *Science* paper. We had said that the time for the dust to fall out of the stratosphere was about three years, which gave it time to spread slowly across the equator; winds would spread it very rapidly across all longitudes, near its original latitude. We based our numbers on the observations we found recorded in a thick volume published by the Royal Society [16] soon after the volcanic explosion of the island of Krakatoa in the Dutch East Indies, in 1883. But at Snowbird, Brian Toon,[5] of the National Aeronautics and Space Administration (Ames, IA), said the dust would fall out in 3–6 months, so our mechanism for getting it from one hemisphere to the other would not work. We therefore were in very serious trouble, except for one comforting fact—we had already seen the iridium layer worldwide, so we knew there had to be a transport mechanism.

How did the Royal Society go wrong, almost a hundred years ago, and how did we recover from that mistake? Professor Stokes, of Stokes's law fame, measured the size of the dust particles by the angular diameter of their diffraction rings, and calculated the time of fallout to be 2–2.5

[5] Toon, O.B., Pollack, J.B., Ackerman, T.P., Turco, R.P., McKay, C.P. & Liu, M.S., Conference on Large Body Impacts and Terrestrial Evolution: Geological, Climatological, and Biological Implications, Snowbird, Utah, Oct. 19–22, 1981.

years, in agreement with the duration of dusty sunsets that were seen worldwide. We took his word for it. We said we thought "our" (much more copious) dust would stay up about three years. But, more recently, dust has been found to fall out much more quickly than that because the dust particles grow by accretion and, as Stokes's equation predicts, fall faster. After Krakatoa, the "dusty sunsets" were at first made by the dust, but it fell out in 3–6 months. Unknown to Professor Stokes, the job of making the sunsets dusty was smoothly taken over by the much finer aerosols that accompany *volcanic* eruptions but not impact explosions. They did their work for the next two years, but Toon correctly pointed out that we could not use such aerosols to keep the sky dark, 65 million years ago.

We knew there had to be a mechanism to get the dust spread worldwide, but our original idea that it was spread through the stratosphere went down the drain. It takes more than one year for material suspended in the atmosphere to move from the northern hemisphere to the southern hemisphere. The Russian hydrogen bomb tests in the 1950s made a lot of carbon-14, and that was observed to move from the northern hemisphere to the southern in about a year. If the dust had fallen out in 3–6 months, it could not have gotten from one hemisphere to the other. But we had already found it in both hemispheres. Something was wrong.

Fortunately, the next day at the Snowbird Conference, two groups reported that the material got spread not by stratospheric winds but by either of two much faster mechanisms. Jones and Kodis[6] from Los Alamos, showed that the material actually went into ballistic orbits and was spread worldwide in a matter of hours. We had known, of course (from a calculation on the back of an envelope), that there was enough energy brought in by the asteroid to put the observed material into ballistic orbit, but we could not think of a detailed mechanism that would accomplish that feat. We did not see how you could get the little particles up through the atmosphere, but people at Los Alamos and Pasadena used very large computers and ran a simulation of an asteroid coming down and hitting the earth. It turned out that convective vertical winds in the fireball did the job. They analyzed a cylindrical asteroid coming downward vertically; the symmetry introduced in this way simplified the calculations. Both groups showed that when the asteroid hit it would distribute the material worldwide, as we saw it distributed very rapidly, and that it would be diluted by between 20 and 100 times its incoming weight, also as we had seen.

---

[6] Jones, E.M. & Kodis, J.W., Los Alamos Report LA-UR-81-3495 and Conference on Large Body Impacts and Terrestrial Evolution: Geological, Climatological, and Biological Implications, Snowbird, Utah, Oct. 19–22, 1981.

All of a sudden, everything was in great shape. The computers did not know that we were in trouble, but they got us out of it very nicely. It turned out that Ahrens and O'Keefe,[7] who did their work in Pasadena, were actually wired in by a special line to the Berkeley computer. That computer was down in the basement of our building, cranking away on this problem of great interest to us, and we did not know it.

Now, for a couple of other odd facts. Miriam Kastner [17, 18] of the Scripps Institute of Oceanography, has shown that the boundary clay layer in Denmark was a glass 65 million years ago as a result of a volcanic or impact eruption. Smit and Klaver [19, 20] found, in Spain and Tunisia, large numbers of very unusual, tiny "sanidine spherules" embedded in a very narrow iridium-bearing clay layer at the C-T boundary. Smit [20] argued from these that the layer is either of impact or volcanic origin and, because the relative abundances of the rare elements match that of carbonaceous chondrites, but not that of crustal or mantle material, he concluded that it is of impact origin. These two separate observations confirm our implied prediction number seven. Alesandro Montanari, a student of Walter's, has also found these same unusual spherules in the Italian clay.

Smit has shown that the sanidine-bearing layer in Spain, where he does his work, is about 1 mm thick, which shows that it was deposited in a period of 50 to 300 years (or less). From this, paleontologists have a time marker which is seen worldwide, and which we now know, from geological observations, to have been laid down in an exceedingly short time. From the computer simulations, which I happen to believe, we know that the layer was laid down even faster. The so-called hydrodynamic computer programs used in these computer simulations are like the ones used to design nuclear weapons; they involve temperatures, pressures, and material velocities much higher than those found under normal conditions, and they are known to do their tasks with great precision. A typical computer run involves many billions of numerical calculations. So far as I know, such great computing power has never before been brought to bear on problems of interest to paleontologists.

Now, as far as killing mechanisms are concerned, we had trouble finding our first one. We had to discard all culprits but the asteroid. Finally, we said, "Okay, let us accept the fact that the material that we see worldwide had to fall down through the atmosphere. We now see that it is a few centimeters thick. Let us take that material and distribute it in the atmosphere in any kind of particles and with any spacing that

[7] O'Keefe, J.D. & Ahrens, T.J., Conference on Large Body Impacts and Terrestrial Evolution: Geological, Climatological, and Biological Implications, Snowbird, Utah, Oct. 19–22, 1981.

you can imagine. It is going to be very, very opaque.'' We originally thought the sky would be black for three years. Now, the number is 3–6 months, and the scenario that we came up with was that the darkness would stop photosynthesis, all the little phytoplankton on the surface of the ocean would die and fall to the bottom, and the food chain for the larger animals in the sea would be disrupted. On the land, the plants would also die. Herbivores and carnivores alike would die of starvation. That was just the first of several killing scenarios. I am confident that it is the only one we need to explain the catastrophic extinctions in the oceans. The lack of sunlight will quickly kill the phytoplankton in the surface layers and, when that base of the food chain is eliminated, most of the life in the sea will be doomed to a relatively quick death.[8] Thierstein, a paleontologist who specializes in microplankton, is comfortable with this scenario,[9] and stated "Darkness is a very good mechanism that could account for the pattern we have [15].'' In fact, the micropaleontologists, most of whom like the asteroid impact theory, are much happier with the 3–6 months of darkness than they were with the original, longer interval.

Historically, the second one [21] is due to Cesare Emiliani, who is a paleontologist, E.B. Kraus, who is an atmospheric modeler, and Gene Shoemaker, to whom I have already referred. They believe that a greenhouse effect caused by the asteroid hitting the ocean and sending up an enormous amount of water vapor would heat the atmosphere and the environment up by as much as 10°C. That does not seem like very much to me, but they assure us that it would kill a great number of the land animals, particularly near the equator, where the fauna are living close to the maximum tolerable temperature.

Then, Toon and his colleagues[5] came up with a third killing mechanism. They reported that their computer simulations show that it first would be very cold for several months. The temperature would go down to about − 18°C for 6–9 months. That would wipe out most of the animals that did not know how to hibernate.

Recently, a fourth killing scenario has come to light. This one, from Professors Lewis and co-workers,[10,11] is that the enormous amount of

---

[8] Milne, D.H. & McKay, C., Conference on Large Body Impacts and Terrestrial Evolution: Geological, Climatological, and Biological Implications, Snowbird, Utah, Oct. 19–22, 1981.
[9] Thierstein, H.R., Conference on Large Body Impacts and Terrestrial Evolution: Geological, Climatological, and Biological Implications, Snowbird, Utah, Oct. 19–22, 1981.
[10] Lewis, J.S. & Watkins, G.H., "Chemical Consequences of Major Impact Events on Earth," preprint, May 18, 1981.
[11] Hartman, H. & Lewis, J.S., "Cretaceous Extinctions: Effects of Acidification of the Surface Layer of the Oceans," preprint, July 23, 1981.

radiant energy in the rising fireball would go through the atmosphere and fix nitrogen to make enormous amounts of nitrogen oxides. It would make acid rain, the rain would fall into the ocean, and the calcium carbonate-based foraminifers would dissolve in the acidified water. I think the chances are that all four of these scenarios had some part in the various extinctions, and it is going to be a life's work for some people to untangle all these things.

Let me now tell you just how much energy was released when the asteroid hit. A trivial calculation shows that it released an energy of about 100 million megatons. A 1-megaton bomb is a big bomb. This is 10^8 of those. The worst nuclear scenario I have ever heard considered is when all 50,000 bombs that we and the Russians own go off pretty much at the same time. The energy released in that case would be less than what we got in the asteroid impact by a factor of about 10^{-4}. Thus, this asteroid impact was the greatest catastrophe in the history of the earth, of which we have any record, and in fact we have a very good record of it.

I will now comment in some detail on the contrary views of the C-T extinction that have been expressed in print, and in many lectures, by my good friend William Clemens, professor of paleontology at Berkeley, who is certainly the most vocal critic of our work. We have a nice arrangement with Bill. For the past 12 weeks, seven or eight of us have spent every Tuesday morning sitting around a table in his conference room—four members of our group, Bill Clemens, one or two of his students, and Dale Russell, who is on sabbatical leave at Berkeley. Dale is a vertebrate paleontologist whose specialty is the study of dinosaurs. He agrees with us that the dinosaurs were wiped out suddenly as a direct result of the asteroid impact, and he further believes that, had the asteroid not hit the earth 65 million years ago, the mammals could not have evolved the way they did. But he believes that intelligent "humanoids" would have evolved in the class of reptiles. He and one of his colleagues are responsible for a set of pictures that purport to show what these two-legged, upright-walking, intelligent creatures might have looked like. And they might have formed their own National Academy and be discussing what would happen to them when one of the asteroids they saw in their telescopes hit the earth.

Our little group has sat around the table for three hours each time and debated our differences and tried to get to understand how the other person was thinking. I do not think this has happened very often across disciplinary lines in science. It is a good way to settle arguments, even though we still have some serious disagreements. But fortunately, we have remained friends throughout our long period of disagreement.

We are indebted to Bill for getting us samples from Montana that show an iridium enhancement in rocks that are close to dinosaur fossils. Carl Orth's group in New Mexico found the first iridium at a continental site, but there were no dinosaurs around there. Bill Clemens collected samples for us in his favorite hunting grounds at Hell Creek in Montana, one of the greatest sites for finding dinosaurs. Frank Asaro and Helen Michel found a large enhancement of iridium, and that is the first experimental evidence that ties the asteroid impact to the extinction of the dinosaurs (Fig. 4.10). I had given a number of talks to physics department colloquia titled "Asteroids and Dinosaurs" before we had any direct connection between the asteroid impact and the dinosaur extinction. You might consider that to be one of our major predictions—that the asteroid impact led directly to the dinosaur extinction. I think the connection is now extraordinarily well established, but will try to explain why Bill

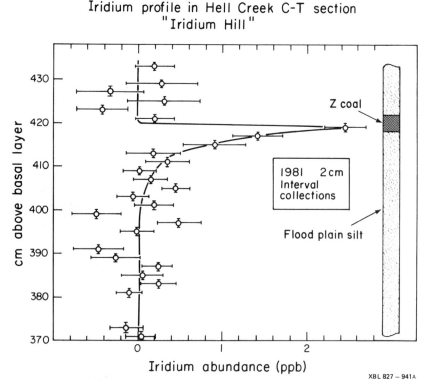

Figure 4.10. Iridium abundance at Iridium Hill section, Hell Creek, Montana, showing the Z coal.

Clemens does not agree with that conclusion, and then will explain why I think his arguments do not stand up under close scrutiny.

Figure 4.11 is Bill Clemens's slide that he has used in a great many talks, and I am indebted to him for letting me use it. He uses this to show that we are wrong in associating the dinosaur extinction with the asteroid impact. The iridium was found in what is called the (basal) "lower Z coal." This coal layer is seen over wide areas in Montana; on this diagram it is shown at the 4.2-meter level. Bill says that the 0.8-meter level is the highest at which he has seen dinosaur bone, and he frequently refers to this as the stratigraphic level at which "the dinosaurs became extinct." (In a recent article [22] with Archibald he says that his student, Lowell Dingus, has seen some dinosaur fossils above the Z coal layer.) Because this is our main point of contention, I will spend some time explaining our differing views concerning the significance of that "highest bone."

Two other features of this very important slide are also worthy of notice, the pollen sample and the fossiliferous zone which Bill usually refers to as a site which produces Paleocene mammal fossils. I will not speak further of the pollen which does not seem to bother Bill nearly

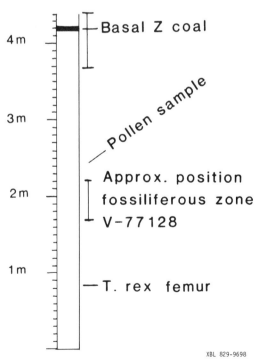

Figure 4.11. Stratigraphic section in Hell Creek, Montana.

XBL 829-9698

as much as the Paleocene (or early Tertiary) mammal fossils do. (These appear at the 2-meter level.) Bill's main interest is in early mammals, rather than in dinosaurs, and he thinks that such mammals have no business being below the iridium layer, if that layer really defines the C-T boundary, which he doubts is the case on the continents—although he is apparently able to accept it in the oceanic sequences. During many of our private discussions, I took the position that evolution does not move fast enough to make the appearance of Paleocene mammal fossils below the iridium layer troublesome to our theory—that the dinosaurs were reproducing at a fairly constant rate, over millions of years, and were suddenly wiped out as a result of the asteroid impact. Paleontologists have never before had such a worldwide sharply defined "horizon" as is furnished by the iridium layer, except for those special cases that happen to coincide with a paleomagnetic reversal. So my argument (in a field in which I have no credentials) was that there was no previous evidence that the Paleocene mammals did not originate 20,000 years before the Paleocene period started—at the C-T boundary.

Now let us look at the time scale that applies to Figure 4.11. In all our long discussions of this figure, the sedimentation rate was assumed by everyone to be about 1 meter in 9000 years. (That average rate comes from the known time between the magnetic reversals that are observed in the Montana sections.) So Bill Clemens is impressed by the fact that the dinosaurs became extinct "long before" the iridium layer was deposited—a difference in height, on this figure, of 3.4 meters or 30,000 years. But the usual description, by paleontologists, of an extinction that took place in the course of 1 million years, is that it "happened rapidly". To someone like me, who is new to the field, it is confusing to hear from the same people that 1 million years is a "short time" and that 30,000 years is a "long time".

In addition to this strange confusion in time scales, I have heard Bill Clemens, and other paleontologists as well, say that the dinosaurs did not disappear suddenly but were declining in population and diversity, all over the world, for a million years or so before they finally became extinct, near the C-T boundary. First of all, I should say that I have looked closely at a lot of data that bear on this alleged "decline", and I agree with Dale Russell that they do not stand up under careful examination. In the last of our 12 seminars to which I have referred, Bill Clemens presented a table of dinosaur fossils that showed that neither the population nor the diversity of dinosaurs had changed appreciably in the 20 meters below the Z coal layer. (At least, that is what the data said to me and to Dale Russell, and Bill Clemens did not attempt to use them to prove otherwise.) Bill's table appropriately showed only "articulated dinosaur fossils'', meaning samples of at least two bones in nearly their formal relationship, or a single bone so large that one could be sure that it had not been shifted from its original site, for instance by running water. There were 17 fossils in the sample, extending downward from the Z coal to a distance of 18.3 meters. This corresponds very nearly to a time interval of 165,000 years. The average spacing was 1.1 meters per dinosaur, and anyone who is used to looking at truly random samples of objects would say, "There is no indication that the population from which this sample was taken was declining as it approached the Z coal layer." It looks extraordinarily uniform to me, even though there is a non-statistically significant *increase* in the number of fossils in the top 30 feet compared to the bottom 30 feet—10 to 7.

I will return to a more detailed discussion of these matters because they are the ones that caused me to come to conclusions quite different from those drawn by Bill Clemens. And I will show that, if Bill Clemens is correct in his "decline hypothesis", it destroys his argument that the eventual extinction of the dinosaurs came before the asteroid impact occurred.

I will now address what I consider to be a serious error in the way Bill Clemens analyzes his data. The field of data analysis is one in which I have had a lot of experience—in contrast to my inexperience in paleontology—so I will offer this criticism without apology. The "*T. rex* femur" that appears at the 0.8-meter level is considered by Bill Clemens to mark the time at which the dinosaurs went extinct. I have "called him" on this point so many times in our little seminars, that I am sure I am not being unfair to him when I say that he really believes that the dinosaurs went extinct 3.4 meters before the iridium layer was deposited, or close to 30,000 years earlier (and in his recent article [22] he made this point several times). The various members of our group have come up with at least four different ways of demonstrating that the proper point to mark the disappearance of the dinosaurs—based on Bill's "highest observed fossil"—is measured in meters (rather than decimeters or decameters) above the "highest bone". (The fact that we proposed several new ways of demonstrating that assertion, on several succeeding Tuesday meetings, is the best proof I can offer that we had not convinced Bill by our earlier arguments. Each week the proposer of the new explanation would say, ahead of time, "I'll bet this one will convince Bill Clemens.")

The easiest way to show what the problem is, and why it is not important in the marine deposits, is to state it in its simplest form. We will assume that some fossils—for example, foraminifers or dinosaurs—are seen in an exposed cliff face, with an average vertical spacing equal to L meters. (If we look at a section of the face of the cliff only half as wide, the appropriate value of L will be twice as large.) The hypothesis we are testing is that the creatures whose fossils we are observing were reproducing at a substantially constant rate until they were suddenly eliminated as the result of some catastrophic event. We have used four separate methods to show that the most probable location of the true "extinction layer" is exactly L meters above the highest observed fossil in that section. The four methods are (*i*) analytical, (*ii*) use of computer-generated plots of randomly occurring "fossils" but with a known cut-off level not indicated on the plot, (*iii*) the Monte Carlo random number method, and (*iv*) an analogy based on locating the United States–Canada border by observing the home of the most northerly US citizen and the home of the northernmost US Congressman. You may enjoy developing this analogy—it works quite well.

The second method corresponds most closely to what one finds in the field—a collection of fossils extending left and right, to the edges of the page, but with no fossils in the upper part of the diagram, above some unmarked line that was at a different height on each page. We passed out dozens of these plots, which were generated on the computer by Walter's student Kevin Stewart at one of our seminars, and asked each participant to guess where the computer had located the "sharp cut-off". In some of these plots, the computer was instructed to weight the surviving fossil population differently in various lithological layers. We did this because Dale Russell's experience as a dinosaur fossil hunter has taught him that there is a larger chance of finding fossils in some formations, such as sandstones, than there is in siltstones or mudstones. So the computer-generated fossil plots corresponded as closely as we could make them to a real field situation. When the "key" was revealed, it was clear that no one had done a good job in locating the iridium layer, but those of us who believed the analytical theorem—that one should pick a point that is above the highest fossil by an amount equal to the average spacing, L—did better than the paleontologists, who have been taught for most of their professional lives to take most seriously the levels corresponding to the "first appearance" and the "last appearance" of any species. The difference between those two levels is called the "range" of the species, and it is accepted that all species do (or will) become extinct at some level.

I believe the reason for the wider acceptance, among paleontologists, of the idea that the asteroid impact led to the extinction of the foraminifers is that the average spacing, L, between their fossils in limestone that crosses the C-T boundary can be a small fraction of a millimeter. The boundary clay has a lower boundary that is definable to only somewhat less than 1 mm, so the coincidence between the iridium layer and the "highest foraminifer" is "perfect", and so "everybody" believes in the causal relationship between the asteroid and the extinction.

In the case of the dinosaur fossils, the average spacing is unknown, but in Bill Clemens's table it is slightly more than 1 meter. If we took it to be exactly 1 meter, and independent of lithological factors, the analytical expression for the chance that the iridium layer appeared at least 3.4 meters above the highest fossil is $p = e^{-3.4} = 0.033$. On the other hand, if the average spacing were 2 meters or 0.5 meter, the probabilities that the iridium layer is where it is are $e^{-1.7}$ or $e^{-6.8}$, equal to 0.183 and 0.0011, respectively. We will soon see that all of these probabilities are larger than the exceedingly small probability that Bill Clemens is forced to accept, when he says that the dinosaurs became extinct, for some unspecified reason unconnected with the asteroid impact he has accepted, about 30,000 years before that impact took place.

It is easy to calculate the probability that the dinosaurs, which had dominated the earth for nearly all of the Mesozoic era—from about 200 million years ago—would become extinct just 30,000 years before any arbitrarily chosen time marker—for example, the asteroid impact; that probability is the ratio of those two times, or 1.5×10^{-4}. As I just said, that is smaller than any of the probabilities we can construct from the "gap" data, and it suffers further from its completely *ad hoc* nature— there is nothing in the history of the earth that can be connected with this extraordinarily coincidental "extinction." On the other hand, our preferred scenario is tied solidly to a well-documented catastrophe that is the most severe event of which we have any record. I really cannot conceal my amazement that some paleontologists prefer to think that the dinosaurs, which had survived all sorts of severe environmental changes and flourished for 140 million years, would suddenly, and for no specified reason, disappear from the face of the earth (to say nothing of the giant reptiles in the oceans and air) in a period measured in tens of thousands of years. I think that if I had spent most of my life studying these admirable and hardy creatures I would have more respect for their tenacity and would argue that they could survive almost any trauma except the worst one that has ever been recorded on earth—the impact of the C-T asteroid.

Because I mentioned the Monte Carlo method of demonstrating that one needs to add L to the height of the highest fossil to locate the most probable position of the iridium layer, I show (Table 4.2) the results of 20 computer-generated dinosaur fossil sequences. Each set was constructed by a random number generator which positioned 50 dinosaur fossils randomly in a stratigraphic height of 100 meters. L is 2.0 meters in all 20 sections. The sharp cut-off at the top is always located at 0 meters

	Level, meters	
Sample	Z	Z + L
1	− 3.201	− 1.201
2	− 3.063	− 1.063
3	− 0.521	+ 1.479
4	− 0.396	+ 1.604
5	− 0.097	+ 1.903
6	− 5.408	− 3.408
7	− 2.930	− 0.930
8	− 0.649	+ 1.351
9	− 3.747	− 1.747
10	− 1.097	+ 0.903
11	− 0.109	+ 1.891
12	− 0.244	+ 1.756
13	− 1.501	+ 0.499
14	− 0.680	+ 1.320
15	− 1.896	+ 0.104
16	− 4.330	− 2.330
17	− 2.903	− 0.903
18	− 3.681	− 1.681
19	− 4.112	− 2.112
20	− 1.665	+ 0.335
Mean	− 2.112	− 0.112

Table 4.2. Computer-generated "highest dinosaur": Monte Carlo table of the highest fossil in 20 random sequences of 50 fossils, each having a density of 50 fossils per 100 meters

L = 2 meters; zero elevation corresponds to true extinction.

and you can see where the highest fossil is located in each section. You also see that, if you assume that the cut-off is at the highest bone, you guess wrong, on the average, but just by L meters. But if you add L meters to the highest bone in each section, then your average estimate of the position of the cut-off in the 20 cases is just right.

I said above that I would point out the trouble Bill Clemens would be in if the "gradual decline" of the dinosaurs turned out to be real—which I continue to doubt. The trouble comes from the fact that the value of L that one must add to the height of the "last observed dinosaur" to locate the "most probable height" of the extinction level is not the *average* value of L observed in some collecting site but the much larger value of L associated with the smaller (declined) population near the time that the "highest fossil" was laid down. Because the probabilities of observing "gaps" ($>G$) between the highest fossil and the iridium layer (assuming it caused the extinction) are equal to $e^{-G/L}$, we see that Bill does not have a statistically significant experimental gap to explain if he really believes in his "decline hypothesis". (The larger L is compared to G, the closer $e^{-G/L}$ approaches unity.)

Two questions that I frequently hear are, "Where did the asteroid hit?" and "How is the theory being accepted these days?" The answer to the first is that we do not know. No crater of the correct size (100–150 km in diameter) and age is known on the earth, with the possible exception of the Deccan Traps region on the Indian subcontinent. Fred Whipple's [23] interesting suggestion that the asteroid hit the mid-Atlantic ridge, between Greenland and Norway, and led to the formation of Iceland unfortunately is wrong because paleomagnetic evidence shows that there was no such ridge at the end of the Cretaceous period—Greenland and Norway had not yet separated. We may never see the crater, because 20% of what was the earth's crust 65 million years ago has since been subducted below the continents. So there is a 20% chance that the crater has disappeared forever, but there is also a finite chance that it still exists on some part of the ocean floor that has not been mapped with sufficient resolution to show it. Many geologists have written to suggest possible impact sites, and each one has looked pretty exciting at first glance. But all of them have had to be discarded, for one reason or another.

I conclude by addressing the question concerning the acceptance of the theory. Almost everyone now believes that a 10-km-diameter asteroid (or comet or meteorite) hit the earth 65 million years ago and wiped out most of the life in the sea. When we first said that the extinctions were caused by an asteroid, we had no information on the detailed composition of the asteroid and, in fact, no one had ever had a chance to analyze an asteroid. But if we had been a little more adventurous, we would have made an eighth prediction—that we would eventually prove that the asteroid had a composition essentially identical to that of the most common solar system debris we know, the carbonaceous chondritic meteorite. We always assumed that it *did* have that composition, but it did not occur to us that there would be a way to prove it. This was first done by Ganapathy [24] who found that the ratios of platinum group elements in the Danish boundary layer corresponded roughly to values in carbonaceous chondritic meteorites. We then measured the Pt/Ir and Au/Ir ratios in Danish and Spanish boundary clays with high precision, and they agreed almost perfectly with type I carbonaceous chondrites. They did not resemble at all crustal or mantle material from anywhere on the earth. The measured ratios also agreed with ratios in two other kinds of chondritic meteorites but not in iron meteorites.

Since Archibald, Clemens, and Hickey all assert that the extinctions were not synchronous—the land plant extinctions, and the land animal (e.g., dinosaur) extinctions—let me end the technical part of my talk with arguments that I find overwhelmingly convincing as to their precise synchroneity.

The Orth graphs, plus the rarity of iridium layers, show that the oceanic and land plant extinctions were synchronous to better than 5 cm, or appreciably less than 1000 years. No data have been presented by any of the three authors I just mentioned that attempt to challenge that conclusion. But they do challenge the simultaneity of the land floral and faunal extinctions, based on the 3-meter "gap" between the "highest dinosaur" and the pollen changes. I cannot think of anything to add to the set of four arguments I have given to show that the "gap" has no experimental significance.

In trying to decide whether we or our critics are correct in our deductions, I suggest comparison of two models. The first is ours, which says the asteroid was responsible for the iridium layer found by Orth in New Mexico and for the ones that we and others have found over the earth and in oceanic sections, and that anyone, using a hand lens, can see was synchronous with the oceanic extinctions. Our model says these two were synchronous to within a few years, so one doesn't need to calculate a probability—the theory simply predicts what we see—simultaneously within the resolution of the observations.

But, if we take the Archibald, Clemens, and Hickey position that the asteroid had nothing to do with the land floral extinctions, then the observed time coincidence of the two events is purely a matter of luck, which can be expressed as a probability. The numerator is the very generous 1000 years I have assigned, and the denominator should be the average time between "spikes" such as the dip in the pollen density. Because I have not heard of other spikes of this nature, I will use for this average time what I think of as the "characteristic species time" or 1 million years. So the probability that the observed simultaneity is due to pure luck unrelated to an asteroid impact is about 10^{-3}. In physics, we do not treat seriously theories with such low *a priori* probabilities. (But if you look closely at the writings of Archibald, Clemens, and Hickey, you find that they do not really have a viable competing theory— one that explains some reasonable fraction of the observational data. I think it is correct to say that their theory is that our theory is wrong!)

The simultaneity of the C-T extinctions in the oceans and on the land can also be demonstrated by a completely different argument, that depends only on foraminifers, dinosaurs, and iridium. Let us look at what our group concluded after finding iridium layers in Italy, Denmark, and New Zealand and deducing that these layers resulted from an asteroid impact.

With the exception of Walter, none of the members of our group knew anything about the extinctions of the land animals. But we were forced to say that there would be an iridium layer seen in continental sites, precisely at the C-T boundary, as defined by the paleontologists. And this prediction relates to dinosaur extinctions on all continents, so we should see iridium layers just above the highest dinosaurs in Western North America, Argentina, France, Spain, and Mongolia. (We have not yet looked at the foreign locations, but I remind you that we did not pick the site to examine; that was a random selection.) Three of us knew nothing about Montana dinosaurs or the lower Z coal layer. But if we had known what Dave and Bill now say about that layer, we would have predicted (number nine) that the iridium enhancement would be found in the lower Z coal layer. (Here is what they say about that layer: "This coal came to represent the Cretaceous-Tertiary boundary in Montana, *because* remains *of dinosaurs* had not been found above it" [emphasis added].) Note that this sentence does not mention pollen or mammals. With no knowledge whatsoever about dinosaurs, we predicted that there should be an iridium enhancement at the (unknown to us) C-T boundary, which Dave and Bill could have told us was in the Z coal layer, and when we looked there, there it was! (Fig. 4.10). Actually, we first looked in the region of Bill Clemens's favored place—three meters below the Z coal layer—and found no "signal". We then worked our way slowly upward, 10 cm at a time, until we found the enhancement.

If you believe the asteroid theory, as we do, then there is nothing surprising about this—that is just where the iridium *had* to be. But if you take the point of view of our paleontologist critics—that the asteroid impact had nothing to do with the dinosaur extinction—then you can calculate the probability that we were simply lucky in that prediction. In this case, the numerator is the thickness of the Z coal layer, or about 4 cm, which we can again approximate as less than 1000 years. The denominator is again undetermined but certainly in the range of millions of years. So my estimate of the probability that we were "lucky", even though our theory was quite invalid, is about 10^{-3}. And, in case you think I'm simply repeating an old argument, I'll remind you that the numerator, 5 cm, in the first probability came from a comparison of the two halves of the Orth graph (Fig. 4.8), whereas the nearly same value for the numerator in the second probability calculation came from the measured thickness of the lower Z coal layer, and our discovery of the iridium enhancement as its base (Fig. 4.10). So the two sets of measurements are quite independent, and the rules of statistics say that we should multiply the two probabilities, to get an obviously absurd chance of the two sets of observations being due to luck; $p = 10^{-6}$. It is also interesting that we did not have to calculate the probability that the iridium layer was in coincidence with the extinction of the foraminifers; that probability has, for its numerator, a distance closer to 1 mm in several places that are widely distributed over the globe.

I hope these exercises will show you why, as an experimentalist, I am convinced that the three extinctions in question were simultaneous—the oceanic extinction, the land floral extinction, and the land faunal extinction.

Before I leave the matter of probabilities, let me remind you that above I calculated the probability that the dinosaurs, which appeared on earth about 200 million years ago, would suddenly become extinct within about three meters, or about 30,000 years of some arbitrarily chosen time marker. (We did the calculation on the assumption that the time marker was the time of the asteroid impact. But if the asteroid had nothing to do with the dinosaur extinction, as our critics believe, then there is no reason to use the asteroid impact as the "arbitrary time marker"—it could in fact be any arbitrarily assigned time.) And, as I showed earlier, the probability that this happened "by luck" was about 1.5×10^{-4}. When I wrote the first draft of this paper, I treated this probability as independent of the other two—its numerator is 30,000 years rather than 1000 years, and its denominator is 200 million years rather than 1 million years. So I multiplied the three probabilities together, to yield an overall probability that all three observations happened by luck—assuming that the asteroid impact had no relationship to either of the land extinctions. But that is probably overkill, for two reasons: I should not use Clemens's erroneous conclusion that the 30,000 year "gap" is significant, to cast further doubt on his gradualistic theory; and the 4-cm limit of error between the Z coal layer and the iridium layer, and the 3-meter interval between the Z coal layer and the "highest dinosaur" are not completely independent; both involve the location of the Z coal layer. But I think that a factor of 10^6 working against the Archibald-Clemens theory is impressive enough.

I conclude this talk with a brief discussion of how a theory is "proved". We all know that theories cannot be proved; they can only be disproved, as Newton's theory of gravitation was disproved by the observations that led to the acceptance of Einstein's theory of gravitation. So let me change my words and ask how theories come to be accepted. Here the classic example is the Copernican heliocentric theory that displaced the Ptolemaic geocentric theory. It became accepted not because Galileo saw the phases of Venus, as most of us believe, but simply because the heliocentric theory easily passed a long series of tests to which it was subjected, whereas to pass those same tests, the geocentric theories had to become more and more contrived. (That is why I've spent so much time telling you of the many tests and predictions that the asteroid theory has "passed".)

Finally, if you feel that I have been too hard on my paleontologist friends and have given the impression that physicists always wear white hats, let me remind you of a time when our greatest physicist, Lord Kelvin, wore a black hat and seriously impeded progress in the earth sciences. We all know that he declared—with no "ifs", "ands", or "buts"—that the geological time scale was all wrong; he was absolutely sure that the sun could not have been shining for more than about 30 million years, using the energy of gravitational collapse.

But most of us do not know that the first man to suggest the answer to this serious problem was Thomas C. Chamberlin [25], a geologist at my alma mater, the University of Chicago. He said that, since the sun had obviously been shining for a much longer time, there must be an as yet undiscovered source of energy in the atoms that make up the sun! And on this occasion, when the tables were turned, the physicists, who had been dragging their heels for a long time, eventually discovered "atomic energy" for themselves (and even convinced everyone that it was "their baby"), and then went on to explain in detail just where the sun's energy comes from.

Every science has much to learn from its sister sciences, and I look forward to the continuation of our cross-disciplinary Tuesday morning sessions.

Appendix

The recent article by Archibald and Clemens [22], "Late Cretaceous Extinctions", is contemporaneous with my talk. Bill Clemens and I had earlier discussed all the points I made in my talk and almost all those made in his new article, so it might be useful for the reader—who might be trying to decide which point of view to adopt—if I comment on a few places where we obviously disagree.

It would make this printed version much too long if I were to address all the points in the article with which I disagree. I will concentrate on the alternative theories that Archibald and Clemens discussed in some detail. They mention explicitly only two such theories, and both can be quickly dismissed. The first—the supernova theory—is not consistent with Orth's limit on the plutonium-244 near a continental boundary; he found less than 10^{-4} of the amount called for by the theory. Furthermore, that theory has already been abandoned by its three chief proponents, Mal Ruderman in physics, Dale Russell in paleontology, and Wallace Tucker in astrophysics.

The second theory is Steve Gartner's "Arctic Spillover Model". This was an acceptable theory when it was proposed, several years ago, but it is no longer acceptable because it offers no reasonable explanation for the iridium layer in the ocean sediments and no possible explanation for the iridium layers found at continental sites. I'm really quite puzzled to see that in 1982 two knowledgeable paleontologists would show such a lack of appreciation for the scientific method as to offer a couple of outmoded theories as their only two alternative theories to that of the asteroid. One can't use the excuse that, when they were proposed, neither could be falsified. The facts of the matter are that as of today, both of them are as dead as the phlogiston theory of chemistry, and I have not heard a serious suggestion in place of the asteroid theory. (But of course that situation has no bearing on whether or not the asteroid theory is correct.)

On the last page of their article, they speak of several vaguely defined noncatastrophic theories, but then they apparently (and I believe correctly) dismiss such theories by saying, "Looking back, it seems unlikely that gradual processes could have caused the extinctions that occurred at the end of the Cretaceous." This evaluation seems to be in good accord with a statement that appears near the beginning of the article, "From today's perspective, the extinction of the dinosaurs some 65 million years ago appears to have occurred almost literally overnight."

After reading their article at least six or eight times, I came away with the feeling that they are emphasizing four main points. First, in field paleontology, it is terribly difficult to make meaningful measurements that tell very much about what happened 65 million years ago. I agree completely with this point, and my admiration for the observations that my newfound friends have made is enormous. But as you can tell, that admiration does not extend to some of the conclusions they draw from those observations.

Their second point is that the dinosaurs disappeared about three meters (approximately 30,000 years) below the C-T boundary. They state this conclusion, explicitly, on four of the eight pages of their article, and it is the point that comes through loudest and clearest. (And you can see that even after trying in four different ways to convince Bill that such a gap has no significance, we really "struck out.")

Their third point is expressed in this way in the article's final sentence, "At present, the admittedly limited, but growing store of data indicates that the biotic changes that occurred before, at, and following the C-T transition were cumulative and gradual and not the result of a single catastrophic event." Again, this point is made on at least four of the eight pages.

Their fourth point is not stated explicitly, but it comes through quite clearly—they do not take seriously the idea that the asteroid impact (if it in fact really occurred, and they never say that they believe that) had anything to do with the extinction of the dinosaurs. There is not a single indication that they take seriously any of the many properties of the iridium layer that I discussed above and which lead me to conclude that the asteroid *did* trigger the dinosaur extinction. (You can be sure that before I make such a sweeping statement, I have carefully read and reread what Dave and Bill said about the iridium layer, each of the 13 times they mentioned the word "iridium".)

It seems to me that their article is in no way responsive to the wealth of data that I have presented in this talk, and with which Dave and Bill are intimately familiar. If George Mallory of Everest fame were still alive, I think he would say, "Gentlemen, you should take the iridium layer seriously—it is there!"

References

[1] Alvarez, L.W., Alvarez, W., Asaro, F., and Michel, H.V. (1980) *Science* **208**, 1095–1108.

[2] Alvarez, W., Alvarez, L.W., Asaro, F., and Michel, H.V. (1979) *EOS* **60**, 734.

[3] Alvarez, W., Alvarez, L.W., Asaro, F., and Michel, H.V. (1979) in *Cretaceous-Tertiary Boundary Events Symposium* (Publisher, Copenhagen), Vol. 2, p. 69.

[4] Alvarez, W., Alvarez, L.W., Asaro, F., and Michel, H.V. (1979) in *Geological Society of America Abstracts with Programs* (Geological Society of America, Washington, DC), Vol. 11, p. 378.

[5] Newell, N.D. (1967) *Geol. Soc. Am. Spec. Pap.* **89**, 63–91.

[6] Raup, D.M., and Sepkowski, J.J., Jr. (1982) *Science* **215**, 1501–1503.

[7] Russell, D.A., and Rice, G., eds. (1982) *K-TEC II, Cretaceous-Tertiary Extinctions and Possible Terrestrial and Extraterrestrial Causes*, Syllogeus No. 39 (National Museum of Natural Sciences, Ottawa, Canada).

[8] Building Knowledge and Understanding—Enduring Assets of Society [Audio Tape 75], American Association for the Advancement of Science Conference, January 1982, Washington, DC.

[9] Alvarez, W., Asaro, F., Michel, H.V., and Alvarez, L.W. (1982) *Science* **216**, 886–888.

[10] Ganapathy, R. (1982) *Science* **216**, 885–886.

[11] Glass, B.P., and Crosbie, J.R. (1982) *Am. Assoc. Pet. Geol. Bull.* **66,** 471–476.
[12] Orth, C.J., Gilmore, J.S., Knight, J.D., Pillmore, C.L., Tschudy, R.H., and Fassett, J.E. (1981) *Science* **214,** 1341–1343.
[13] Hickey, L.J. (1981) *Nature (London)* **292,** 529–531.
[14] Kerr, R.A. (1980) *Science* **210,** 514.
[15] Kerr, R.A. (1981) *Science* **214,** 896.
[16] Symons, G.J., ed. (1888) *The Eruption of Krakatoa and Subsequent Phenomena* (Report of the Krakatoa Committee of the Royal Society, Harrison, London).
[17] Alvarez, W., Asaro, F., Alvarez, L.W., Michel, H.V., Arthur, M.A., Dean, W.E., Johnson, D.A., Kastner, M., Maurasse, F., Revelle, R.R., and Russell, D.A. (1982) in *Terminal Cretaceous Extinction: A Comparative Assessment of Causes* (American Association for the Advancement of Science, Washington, DC) (abstr.).
[18] Russell, D.A., and Rice, G., eds. (1982) *K-TEC II, Cretaceous-Tertiary Extinctions and Possible Terrestrial and Extraterrestrial Causes,* Syllogeus No. 39 (National Museum of Natural Sciences, Ottawa, Canada), p. 8.
[19] Smit, J., and Klaver, G. (1981) *Nature (London)* **292,** 47–49.
[20] Smit, J. (1981) Dissertation (University of Amsterdam).
[21] Emiliani, C., Krause, E.B., and Shoemaker, E.M. (1981) *Earth Planet. Sci. Lett.* **55,** 317–334.
[22] Archibald, J.D., and Clemens, W.A. (1982) *Am. Sci.* **70,** 377–385.
[23] Whipple, F.L. (1980) Center for Astrophysics reprint no. 1384.
[24] Ganapathy, R. (1980) *Science* **209,** 921–923.
[25] Press, F., and Siever, R. (1978) *Earth* (Freeman, San Francisco), 2nd Ed.

5   The Lunar Laboratory

Edward Teller

Thinking about a lunar colony has occupied many fine science fiction writers, including H.G. Wells, for a considerable length of time. I too have been interested in the idea, I hope, on a realistic basis. Today, a lunar laboratory seems to make sense on scientific, technical, and even economic grounds. I have not proceeded beyond general estimates and a few specific points. The details serve partly for illustration and partly to clarify a few differences from generally discussed ideas.

What form should a lunar laboratory take, and what projects should be attempted? In the initial stage, only a minilaboratory could be considered. Such a laboratory would be staffed by about a dozen people. A few months after Sputnik, I was asked an interesting question about space during a Senate hearing: Should there be any female astronauts? I answered that all astronauts should be women—they weigh less and have more sense. Intelligence seems to be better packaged in women. But nowadays, with affirmative action measures, I have to modify my recommendation. There should be six women and six men to staff the first minilaboratory.

The workers would be rotated back to earth after a limited number of months on the moon. Spending an extended length of time in a region of low gravity leads to decalcification of the bones. On the surface of the moon, gravitational acceleration is one-sixth that of the earth. The

NEW DIRECTIONS IN PHYSICS
The Los Alamos 40th Anniversary Volume
ISBN 0-12-492155-8

lunar laboratory people would have to work inside space suits. Whether a person's bones carry less mass and more gravitational acceleration, or more mass (in the form of a space suit) and less gravitational acceleration may not make much difference in terms of stresses on their skeletons. Probably, people can work on the moon for longer than three months without incurring physiological problems. The duration of their stay might last a year. The real limitation may well be psychological.

Locating the living quarters of the lunar colony in a crater has advantages. Because of the periodic intense radiation produced by flaring sunspots, laboratory workers will need considerable, readily available shielding. By building the laboratory over the edge of a crater, the workers would have a quick means of ducking into a shadow. More than one crater may be needed because any single location will be exposed to heat or cold for too long a time.

It would be even more ideal to place the living quarters inside a cave in the wall of a crater. The temperature in lunar caves is more moderate because an average of the lunar day and night temperatures prevails inside a cave. Living quarters in a cave would save energy and provide extra insurance against unusual levels of solar radiation.

Further advantages are gained by placing the base near one of the lunar poles. Both sun (and heat) and shadow (and extreme cold) would be within easy access. Similarly, from such a point, the earth can be seen, which is useful for communications. Yet nearby areas are shadowed from the earth, which is a preferable situation for astronomical observations.

There seem to be three appropriate craters close to the south pole. Craters are scarcer near the north pole. Starting from the south pole during the safe period of sunspot minima, when no solar flares are expected, the workers could spread out and establish various projects.

A great deal of material will be required if a dozen workers are to work effectively on the moon. My estimate is that a minimum of twenty tons of material per person will be required each year. Only a small fraction of that weight will have to do with supplying the worker's physical needs for food and water. Obviously, their water would be recycled, a technology already well developed. The main fraction of the weight will go for energy (crucial to their survival as well as their work) and equipment for building and research.

The suggestion has been made that we learn to grow food on the moon so that the colony could be self-sustaining. That attempt should be postponed. The valuable opportunity of being on the moon seems wasted in mastering agricultural activities. Man's development on earth has been from hunter to farmer to technologist. On the moon, the first act should be technology. Agriculture can wait until the United Lunar

Colonies formulate their Declaration of Independence. Hunters on the moon will be out of luck.

The main expense of the lunar laboratory is apt to be the transportation of material to the moon. That expense is what caused me to set the size of the lunar colony at twelve people. Flying the shuttle costs much more than was originally expected: Each trip carrying 30 tons of payload costs about $70 million. Considering the additional rocket fuel required to carry such a payload all the way to the moon, the cost may easily triple.

At a cost of $200 million for 30 tons, the annual expense of transporting 240 tons to the lunar laboratory (12 people times 20 tons apiece) would be $1.6 billion. That figure does not include the cost of preparatory research and fabrication of the materials to be delivered. The development of the necessary technology (including transfer vehicles) will take about three years and several billion dollars. But combining all the expenses (including transportation of the lunar laboratory workers), a total annual budget of $3 billion might suffice. Remarkably enough, this is not a great sum in comparison with past NASA expenditures.

Great scientific and industrial benefits can be expected from a lunar laboratory, but the first priority should be to create a station to refuel rockets with oxygen and possibly hydrogen. Rockets can be accelerated to a velocity of approximately 4 kilometers per second (km/sec) by an amount of fuel comparable to the weight of the payload. Each additional 4 km/sec acceleration requires a doubling of the fuel. Thus, leaving the earth (which requires 11 km/sec) plus landing on the moon requires a great amount of fuel; further maneuvers that require fuel become very expensive. Takeoff from the moon (which requires 2.4 km/sec) and orbital velocity around the moon (1.5 km/sec) are comparatively cheap if lunar-produced fuel becomes available.

Even if a few years were required for its establishment, the benefits of a refueling station outweigh all other advantages. It would make the lunar laboratory much more practical by making commuting back and forth cheaper. Lunar refueling will postpone or eliminate the need for nuclear-fueled rockets. The refueling station is the real basis on which the future of our space enterprise depends.

The specific importance of the moon is that it contains plenty of "green cheese" that can be turned into useful products. Fuel for space travel could be obtained from either of two sources. Lunar rocks are essentially oxides. Those oxides should be selected from which oxygen is most easily liberated. Mechanical energy will be needed to crush the rocks so that the oxygen can escape more easily.

An alternative to using lunar rock is to use lunar dust (regolith) since it already has the proper physical form. Regolith consists of particles one millimeter to one micron in size and covers most of the lunar surface. However, its chemical composition makes it a less desirable source of oxygen, and mechanical means will be needed to collect sufficient quantities. On the other hand, regolith is a good source of hydrogen, present at a concentration of 10^{19} atoms per cubic centimeter (less than one atom in a thousand) in a lightly bound form. The hydrogen has been deposited by the solar wind. Thus, fuel for a hydrogen-oxygen rocket could be made available from lunar material. Even if only oxygen is made available, large savings will be made. Transporting hydrogen from the earth involves considerably less weight than transporting oxygen.

Liquefaction of the fuel will not be a problem, because the lunar nighttime temperature is sufficient to accomplish the process. In the case of hydrogen, a little more effort is needed for liquefaction. However, obtaining the fuel will require great amounts of energy. In addition, the lunar colony will need some energy to conduct its work as well as to survive. Massive amounts of energy for fuel production and a far smaller quantity for general use are needed. Two sources seem obvious—solar energy and nuclear energy.

One method of heating the lunar rock would be to focus solar light on a small area. Temperatures up to 3000° Kelvin—which are certainly sufficient—can be achieved. The difficulty in this approach is that the system of mirrors required to reach those temperatures is apt to be heavier than a reactor and, therefore, more expensive. The question is not whether fuel can be produced using solar energy but how much it would cost.

One advantage of solar energy, particularly solar electricity, is that it can be made available in widely distributed locations on the moon. However, if solar cells are used, the energy supply will disappear during the 14 days of solar night, which occur to a lesser extent and in a different way near the poles. Maintaining a continuous supply of energy will require batteries, which are heavy and therefore expensive to transport. Ultimately, ways may be found to make batteries out of lunar materials.

One suggestion that arose a few years ago in connection with solar energy was that it should be collected on a satellite, converted into microwaves, and beamed back to earth for reconversion into electricity. The idea seems unlikely to become economically feasible. But if practicality can be approached, the moon would be a better location for the initial conversion of solar energy.

Using a thermoelectric source based on the temperature differences on the surface and a few feet below the surface should also be considered. Although the sign of the difference changes, the temperature difference is available day and night.

A nuclear reactor would provide a good source of major amounts of heat. Transporting the heavy shield for a reactor would be unnecessary. Only a specially constructed core of a nuclear reactor would have to be sent. The shielding material could be made of lunar rock. In fact, the whole reactor could be built into a lunar cavity in such a way that it would be well shielded. Lightweight excavation equipment will be essential for that task as well as for several other purposes.

Cooling and maintaining a lunar nuclear reactor present problems. Therefore, if a nuclear reactor is used at all, it should be specially constructed in a simple manner, mainly for use in obtaining the fuel for rockets. The reactor must be safe and sturdy; in case of malfunction, it should be replaced rather than repaired. Those questions, of course, will require research.

A further requirement is to keep the oxygen that is produced free of radioactive contamination. Oxygen itself will not give rise to disturbing long-lived radioactivity, but materials associated with the oxygen do. Methods would have to be developed to eliminate traces of them. While in principle this should be possible, carrying it out with remote-control apparatus would not be easy. Experiments to develop appropriate processes would have to be carried out beforehand on earth using lunar materials.

Ultimately, however, both solar and nuclear sources of energy should be available. The nuclear source could provide massive amounts of heat; the solar source would offer a modest amount of electricity in the areas where special projects might be located.

As a refueling center, the moon could serve as a jumping-off station for the exploration of the whole planetary system. One of the great advantages of a lunar colony is that it would make the whole space program considerably less expensive. Having to bring all fuel from earth and having to carry the fuel to overcome solar gravity is extremely expensive. Refueling on the moon would lead to dramatic savings. In the twenty-first century, laboratories on small man-made moons, on moons of other planets, and eventually on planets themselves might be established.

One cannot say that the refueling function of the lunar colony would pay for itself, because exploring the solar system may give us nothing except knowledge. That knowledge may provide us with enormous advantages, but knowledge—unless we regard it as intellectual energy (corresponding to infinitesimal fractions of micrograms) and invoke the $E = mc^2$ equation—has no weight.

After establishing the refueling station, the next task I would propose is not to go on to the solar system but to explore the moon itself. We know almost nothing about the geology of the moon, a branch of science known as selenology. A little is known about what is a few feet under the surface, and the rest is inference. Obtaining cores down to a few thousand feet would help explore the history of the moon. Taking corings at great depths may turn out to be somewhat expensive and will hardly be feasible during this century.

Knowledge about crater formation would also be gained. Most craters were probably made by meteoric impact, but a few may have been made by volcanic eruption. Furthermore, those two phenomena are probably not entirely independent. A large meteor impact may well have effects on the lava layers that, in some time sequence, show up as volcanic action. The relationship between the maria, the flat "seas" of the moon, and the highlands, which are full of craters, is incompletely known. The surface mapping that a colony could carry out would be a great improvement over our present knowledge.

Perhaps the most unexpected advantage of a lunar laboratory may be economic. Having spoken of $3 billion in expense, this may seem improbable. However, remarkable possibilities in pure research as well as economies in industrial applications and defense make the enterprise appear promising.

One vital need in avoiding a nuclear war is to know when any rockets are launched on earth. The best location for observation satellites—our eyes in the sky—is in synchronous orbit at about seven earth radii above us. The question of how to defend this extraordinarily vital link in our defense system is most difficult. The numerous Soviet satellites on elliptical Molniya orbits appear to be more durable.

One expensive but effective proposal for defending observation satellites, from laser attack and from x-rays produced by a nuclear explosion, is to put a heavy shield—a lot of mass—around them. No matter what the material is, its mass would be useful. However, transportation of such mass is expensive.

That strategic location—the synchronous orbit—is less expensive to reach from the moon than from the earth. As mentioned, the velocity change needed to get into synchronous orbit starting from the moon is one-fourth of the velocity change required for a start from earth. Starting from the moon, the needed fuel weighs less than the payload, whereas starting from earth, approximately ten times as much fuel as payload is needed to get into space. Thus, if energy on the moon becomes available and the project is undertaken from the moon using lunar rocks, the expense of putting protective materials around satellites could be greatly decreased.

Industrial applications also may prove valuable. Some time ago, I proposed that NASA should adopt as its theme song, "I've got plenty of nothing, and nothing's plenty for me." Indeed, an obvious use for the lunar laboratory is connected with nothing—that is, a cheap and excellent vacuum. What sort of vacuum does the moon have? At this time, we cannot know accurately, because the astronauts contaminate their immediate surroundings. The lunar colony itself will contaminate the vacuum, but I suspect that it will not have an appreciable effect. The moon itself is emitting gases. Intense ultraviolet solar radiation, micrometeoric impacts, large meteor impacts, and volcanic events all disturb the vacuum.

However, the escape velocity from the moon is so small that the lunar noontime temperature is sufficient for hydrogen to escape. Heavier gases escape more slowly. The atmosphere of the moon—whatever there is of it—rotates with the moon. During the fourteen days of lunar night, it experiences extreme cold and condenses. With the sunrise, the gas again evaporates, rises, and diffuses to the dark side, which acts as a trap. The lunar motion sweeps the atmosphere toward sunrise, and the sun pushes the atmosphere back into the presunrise area. Whatever gas is present is concentrated on one moving longitude—a longitude around the dark edge of sunrise.

Measurements of the moon's atmosphere should be conducted in this area. The pole itself will be quite interesting, because some material might accumulate in its permanently shadowed regions. This is the most likely place for water to be found. The discovery of water deposits would, of course, change many considerations. Then the best way to obtain fuel for propulsion could be through electrolyzing water, using a substantial nuclear reactor or an appropriate solar source of electricity. Thus far, however, the search for water on the moon has proved futile.

Discovering the quality of the vacuum on the moon will have to be done by remote-controlled experiments. The search for the best vacuum could be conducted in the lunar night. Should the moon prove to have an excellent vacuum (which is probable), it could lead to one really important application: Surface chemistry could develop from an art into a science.

On earth, breaking a solid into two parts is an irreversible process. The basic reason is that breaking occurs in an irregular fashion and the parts no longer fit together. However, even when the parts do fit—for example, in the case of carefully broken graphite or mica, which come apart in molecularly plane surfaces—the flat surfaces cannot be made to adhere again. The reason is that before the two pieces can be brought into contact, a monomolecular layer of impurities is deposited from the surrounding gas onto the sheared surface; this destroys the possibility for fully effective adhesion.

The moon may possess a vacuum approaching that of interplanetary space—less than 1000 molecules per cubic centimeter. Registering the degree to which mica adheres after being broken and put back together might be an extremely primitive but effective way to detect an excellent vacuum.

Surface chemistry is of obviously great importance in electronics. One of the possibilities that would enable a moon laboratory to pay for itself is that the availability of an excellent vacuum might lift electronics to an entirely new state of perfection. Far more effective chips for computers might first be produced on the moon. The first step toward making such a machine would be to learn more about surfaces.

Some of the early projects for the lunar base would involve making astronomical observations. Mirrors may be composed of small plane elements, adjusted electronically. I suspect that such a telescope placed on the moon would be quite effective. One advantage is that mirrors on the moon can be completely shielded from the earth and the sun. Moon-quakes tend generally to be much smaller than earthquakes and would present no difficulties for adjustable mirrors. The main difficulty is that the temperature change between lunar day and night will necessitate careful construction and readjustment of the mirrors unless the mirror is located in a permanently shadowed region near the pole. The main competitor of a lunar mirror is, of course, a space mirror.

The most exciting aspect of lunar and space observations is that they are not limited to visible light, in which the wavelength varies by hardly more than a factor of two. All wavelengths—gamma, x-ray, ultraviolet, infrared, and all radar and radio emissions—are easily observed from the moon. Observations of the sun would also improve because both the atmospheric interference and the perturbation from the noise originating in our radio emissions would be eliminated.

Astronomy is the most ancient science. It continues to hold public interest. The amount of funding this branch of science has received has been quite limited, yet the recent progress made in astronomy is at least as great as that in any other part of physics. For example, quasars, pulsars, and neutron stars are beautiful and radically new discoveries. The lunar colony, by putting some effort into a study of our own galaxy and of other galaxies, might well improve our knowledge of the origin of the universe. Such basic information is scarce, and the difficulties of finding out more are challenging.

The lunar laboratory may also enrich the study of high-energy physics. At present, cosmic rays that have exceedingly high energies are used for exploratory purposes in a limited way. By the time the most energetic particles reach the earth's surface, they are contaminated by interaction with the atmosphere. The best way to begin collecting observations from particles having 1000 GeV or more might be to dig long collimating holes on the moon. While the frequency of events would be low, the chance of seeing interesting events is a certainty.

One final high-energy physics project deserves consideration: putting an accelerator on the moon. It would consist of separate accelerating units of limited dimensions with long free run for the particles between them. Because of the small size of lunar quakes and the general cleanliness of the environment, such an accelerator is a real possibility. The problem would be to construct the elements that deflect the particles. If this can be done using lunar materials—for instance, to construct the deflecting magnets—a wonderful accelerator could be constructed. It is easy to imagine such an accelerator on the rim of a big crater.

This suggestion allows me to talk about one man who lives on in the minds and hearts of many of us: Enrico Fermi. Fermi, it is said, never showed a slide in his life, and I have tried to emulate that practice. However, the statement about Fermi is slightly exaggerated. He did show one slide, and that was during a talk he gave on accelerators and their probable development. Fermi's slide showed an accelerator encircling the earth. I do not believe that this will ever become reality on earth, but our colony on the moon might conceivably complete a smaller version around the earth's satellite.

Forty years ago, Robert Oppenheimer talked about having a couple of hundred people working at the weapons laboratory at Los Alamos. By the end of the war, Los Alamos had a population of ten thousand. The lunar base, unfortunately, cannot expand similarly in the near future—the GNP will not allow it. However, I might dare to hope that the lunar laboratory will have twelve people on the five-hundredth anniversary of Columbus' first trip and one hundred in ten more years. Dr. Hans Mark, the deputy administrator of NASA, has estimated, in analogy with the population of Antarctica, that by the year 2030 the lunar colony may have 10,000 people.

Even with the development of refueling on the moon, the expense of such a colony might approach a trillion dollars a year. Would such an effort be reasonable? In view of the great and varied potential benefits, of which we can now see only the bare beginnings, I would not hesitate to wish it so. We must remember that what we can imagine today may be dwarfed by future realities. There are limits to our imagination. There are hardly any to the developments that the human mind and human activity can accomplish.

6   The Future of Particle Accelerators: Post-WWII and Now

R.R. Wilson

In 1945 at the end of WWII, nuclear and particle physicists were at a crossroads. They are again at a crossroads. What happened as physicists contemplated an uncertain future in 1945 and what is happening today as they contemplate an even less certain future?

Physicists were perplexed as they returned to their universities after the war; their perspectives had been changed about the size and cost of the facilities that might be appropriate for them to have in their post-war laboratories. Before the war, Ernest Lawrence had established the Berkeley Radiation Laboratory and a scale of doing things in a big way that had been emulated, even exceeded, in the war-time laboratories which had been created for weapons research—the MIT Radiation Laboratory for radar and Los Alamos for atomic bombs being two of many such labs. But was it physics? Still the physicists who had worked in those labs had learned how much could be accomplished when the support for their experiments was much larger than that to which they had been previously accustomed, and that, typically, had been a few thousand dollars a year. Livingston tells us that even at the "old Rad Lab" he spent only about $1000 for the first cyclotron. During the war physicists had become used to spending millions and working with large teams of associates.

NEW DIRECTIONS IN PHYSICS
The Los Alamos 40th Anniversary Volume
ISBN 0-12-492155-8

The question was: Would this new point of view carry over to the university environment when they returned? And if the spending would really be greater, where would the money be found? Before the war, there had been a strong tradition that all the funds for research at universities be from private sources, else how, indeed, could research be free? The problem came down to this then, whether to succumb to Berkeleyitis, or to be pure but poor.

The problem was, of course, resolved in various ways, but the story of that resolution at Harvard is telling. When the cyclotron had been borrowed from Harvard in 1943 for use at Los Alamos on a rent-free basis, a commitment had been made by the Manhattan Project that it would be returned at the end of the war. When that time came, Ken Bainbridge, who had built it, and I, a new Harvard faculty member, decided that what we wanted was to have it returned. We were both tired—tired of high-pressure war research, tired of administration, tired of bureaucracy. We wanted to go back to our previous halcyon lives of teaching and research.

Our decision to hold General Groves to his commitment brought forth anguished screams from our colleagues who had elected to stay on at Los Alamos. They needed that cyclotron to be an integral part of the post-war laboratory they envisioned building, and they strongly suggested that we leave it at Los Alamos and build a copy at Harvard. "No", we said emphatically, "we're tired of building things for the army". "How about your patriotic duty?" they asked. That kind appeal worked during the war, but not after. We remained adamant. Oppy intervened; a polite "No" in response. General Groves interceded with an offer of money. "No, never! 'Take away your billion dollars, we'll be physicists again'", we happily sang à la Art Roberts.

Ernest Lawrence then placed an almost Faustian temptation before us. If we would leave the cyclotron at Los Alamos, the Radiation Lab would design at Berkeley and help build a new synchrocyclotron (just invented at Los Alamos by Ed McMillan) that would give ten times the energy of the old cyclotron. Furthermore, Lawrence would help find the money with which to build it. In the face of this—plus the censure of our friends—we weakened, then gave in. I believe that this may have been the first accelerator (not counting those at Berkeley) to have been helped substantially by a government grant—in this case by the Office of Naval Research, a progenitor of the NSF. Having blazed that trail, it was followed at other universities by contracts for the construction and use of one after another accelerator, each larger than the preceding one. The series of about a dozen culminated in the 400 MeV Chicago cyclotron and the 450 MeV Columbia cyclotron.

Perhaps we at Harvard suffered by being first for the energy of our protons, 100 MeV, was on the low side—too low to make pions, for example. On the other hand, it was built sooner than the others. Mostly what we wanted at the time was just to get back to doing physics as soon as possible. It is ironic that (again with the exception of the synchrocyclotron at Berkeley) all of that series of synchrocyclotrons have now been dismantled except for the Harvard machine. It had an ideal energy for treating cancer tumors with the proton beam, and it is still busily employed doing just that.

In any case, accelerator laboratories were engendered at dozens of universities, and physics flourished. Nuclear and particle physics were almost an American monopoly for the next decade—perhaps even for the decade that followed—in part because the physicists had not returned to their old pre-war parsimonious ways, but in larger part because visionaries in Washington created new procedures by which government funds were made available for university research without stifling conditions or excessive bureaucracy.

It is not my intention to describe the Golden Age of Physics that followed this largess—indeed, it is still going on—or even to describe the accelerators which in large measure made it possible. Instead, let me just encapsulate the remarkable development of accelerators by referring to the "Livingston chart" in which the log of the particle energy is plotted against the date of attaining that energy. Typically, a particular kind of accelerator shows increases in energy exponentially for several years after its invention then seems to approach an asymptotic limit. Remarkable in the plot is that as one type of accelerator nears its limit another kind of accelerator is invented and continues the upward climb until yet a different technique takes over and carries the torch of energy even higher. The envelope of these different curves, by some curious accident, turns out to be quite linear—a factor of ten increase in energy occurring in roughly seven years. This adds up to a factor of about 100 million increase during those Fifty Golden Years. If we think of the accelerator and its associated detectors as a microscope, then the resolving power has increased by a factor of over ten thousand. Thus, from the nuclear dimension of about 10^{-12} cm which we could just make out in 1930, we can now begin to see down to 10^{-16} cm. And what exciting worlds we have seen: from the wrongly conceived, but magnificently simple, world of the twenties, made only of protons and electrons, to the complex world made up of the hundreds of "elementary" particles of the sixties, to the again relatively simple world of quarks and leptons of today—and with the electric and weak force unified, and with QED and then QCD theories to explain it all. What thrilling progress! What a time to have lived through! Oh, we happy, happy few! More realistically, many, even most, of the old fundamental problems remain to be solved, the masses of the proton, electron, muon, etc., and the unification of all the forces.

But what of today? ... and tomorrow? Today, although particle physics is still robust in the US, we seem to have been outclassed for several years by our European colleagues. For some time they have been spending more than twice as much on this field as we in the US do, their activity being largely centered at CERN, the Western European international laboratory, and at DESY, the German national laboratory at Hamburg.

Money isn't everything, of course. What is serious is that the Europeans have learned the American "can do" approach to physics and have forged ahead of us considerably. Their PETRA colliding electron facility came into operation several years earlier than a similar facility, PEP, at Stanford. The transformation of CERN's 400 GeV proton synchrotron into a proton-antiproton colliding beam facility that has been operating for almost two years is another demonstration of their skill. The demonstration of gluons at PETRA and the discovery of the weak mediating bosons at CERN have been breathtakingly admirable discoveries.

I must confess to a certain ambivalence with regard to these important discoveries. Mainly, I realize that physics in essence is international and that in a few years the place where a discovery has been made becomes almost completely irrelevant as physics becomes formulated and then reformulated. In that spirit, I can enthuse just as much about CERN's contributions as I could had they been made here—and I do enthuse!

On the other hand, the patriot in me recognizes that it does matter whether we contribute vigorously to science or whether we just read about what other nations contribute. It matters because there is a relationship between what we accomplish in any field and our pride in ourselves as Americans, our culture, our education, our technology ... and our well-being. Discoveries, both theoretical and experimental, are a significant measure of the vigor and health of a science—perhaps the only measure. It is this vigor which informs the educational process, which leads us on to new science, new art, new technology, and new industry.

Our present accelerators are respectable, but by no means in advance of the European accelerators. The PEP project of colliding electron beams is now in excellent operation although at a considerably lower energy than its competitor, PETRA. The Tevatron at Fermilab is now coming into operation and we should expect to see 1 TeV (1000 GeV) protons available for experiments there before long. A few years from now it will have been improved so that it can also serve as a proton-antiproton colliding beam facility, giving 2 TeV in the center of mass as compared to the roughly 0.5 TeV available now at CERN (but they will increase their energy, too). For secondary beams of neutrinos and various kinds of mesons we in the US should be more than competitive at Fermilab for several years down the line, and eventually we should have a window of advantage with the colliding hadron beams at Fermilab.

An imaginative project at SLAC is under way to construct a single-pass colliding electron beam facility (SLC). This is sometimes referred to as a "Z^0 Factory" because it will give about 100 GeV c.m. (center of mass), enough to make the Z^0 intermediate boson, the existence of which has just been confirmed at CERN. The SLAC project is somewhat overshadowed by the more conventional LEP project at CERN which is also underway and which should eventually reach twice the energy in electron-positron collisions, should have several collision points, and should provide for a substantially larger collision rate (luminosity). It will also be possible to place superconducting magnets in the large tunnel of LEP, about ten miles in diameter, to produce protons of about 10 TeV in energy.

At Hamburg, a large project, HERA, is underway to build a colliding beam facility in which electron-proton collisions can be studied. This will be comprised of a 840 GeV superconducting proton ring quite similar to that of the Tevatron, and a 30 GeV intersecting electron ring. This should be a superb facility for exploring an unknown part of nature. Unfortunately, we have nothing yet on our books to compare with it.

Soviet physicists are building a 3 TeV proton synchrotron, the UNK project at Serpuchov. It will use their 76 GeV machine as an injector into a ring that will initially have a 400 GeV conventional synchrotron which will in turn serve as injector for the superconducting 3 GeV machine. They anticipate having collisions between the beams in the two rings (2.3 TeV c.m.) and later on having proton-antiproton collisions in the 3 TeV ring (6 TeV c.m.).

In Japan is the TRISTAN project. The first stage is to build a 30 GeV electron-positron colliding beam. Later they expect to have a 200 GeV proton ring in order to have an electron-proton colliding beam facility.

From all this, one can gauge the seriousness of the effort being made on an international level. Beyond the Tevatron and the SPC we have in the US the Colliding Beam Accelerator (CBA), a new name for the old Isabelle project at Brookhaven National Laboratory. It is a proton-proton colliding facility with two 400 GeV beams that should have high luminosity—if it is built, a matter in contention as of this date (April 1983). At Fermilab a new project, the so-called Dedicated Collider, has just been put forward. A large site-filling ring is proposed which should give proton-antiproton collisions of 4 or 5 TeV cm energy. In addition to the obvious advantage of energy that it provides, it would have several interaction points, would provide an increase in luminosity of as much as a factor of one thousand over the Tevatron in its colliding mode. By the addition of an electron ring, it could outclass HERA. There are strong proponents of both of these projects, and there is much that is admirable about both of them, but even if both should be built, it will not be enough to restore the preeminent position in particle physics that we once enjoyed.

Many of us who have been worried about the health of US particle physics believe we should take a larger step beyond the CBA and the Dedicated Collider, with as little delay as possible. What we have proposed as a significant step is a large proton synchrotron-collider that will produce 40 or even 50 TeV c.m. Such a machine might have a diameter of as much as 50 miles, depending on the magnetic field of the magnets. This possibility, sometimes referred to as the "Desertron", was studied last summer at a workshop sponsored by the APS Division of Particle Physics at Snowmass, Colorado. There the Desertron was judged to be desirable and feasible to build. It has also been considered at a workshop of accelerator experts which met at Cornell University for the week of March 28 to April 1(!), 1983. They agreed that it was no joke. Building such an accelerator was within the present technology now, they affirmed, and that it might cost about 2.5 ± 0.5 billion dollars if built today. They also agreed that with a few years of research and development it might cost significantly less than that. Interestingly enough, the magnetic field in the magnets does not seem to make much difference in the cost of the machine. A magnetic field of 2.5 tesla in which the field is shaped by iron but excited by current in a superconductor ("superferric") would require a very large diameter of relatively cheap magnets, while the more standard 5 tesla superconductive magnets, in which only a small part or none of the field comes from the iron, would require half the diameter and half the number of magnets, but each magnet, by present technology, would be more than twice as expensive. Similarly, 10 tesla magnets, which are at the edge of, or slightly beyond, present technology, would be even more expensive. The low field magnets have the advantage that we could build to 20 TeV today and add higher field magnets for higher energy tomorrow—just as the Tevatron has grown out of the original 400 GeV machine.

Although the ring will necessarily have to be considerably larger than the ring at Fermilab, the site itself may be of comparable dimensions. The ring might be considered as a passive element, once it has been fabricated, so that only the underground rights, or rights to a narrow swath just over the ring, need be acquired. The tunnel might be a buried pipe similar to a large gas pipe, and then robots might do the installation, adjustment, and repair of the ring. The site would then contain the collision points, placed rather close together either in series or on parallel branches of the ring. The site would also contain the injectors and the usual collection of shops, offices, etc.

Now, quite conceivably, we in the US may not yet be ready to support the expense of such a project. That is a situation that many particle physicists have foreseen as occurring should the level of expense become too great for one country to bear. What we had envisioned was that, when the fateful day arrived, we would be ready with an alternative in which the cost of a supranational accelerator would be shared between many, hopefully all, of the nations of the world—just as CERN was formed after WWII when no single nation of Western Europe could afford to compete with the US explosion of nuclear facilities I've already discussed.

This possibility first surfaced when, under Eisenhower's Atoms for Peace program, teams of Soviet and US scientists were exchanged in several fields. I was a member of the Particle Physics team and remember that a part of our instructions had been to explore the possibility of jointly building a large accelerator. The downing of an ill-fated US spy plane over Russia just as our team went over made that particular idea moot.

In 1974, an international meeting of all laboratory heads and other appropriate people was held in New Orleans under AEC auspices. A world laboratory to build a great accelerator was one of the principal ideas discussed there. That led to the formation of ICFA (the International Committee for Future Accelerators) under IUPAP auspices. ICFA has arranged workshops in the US, in Switzerland and in the USSR. Indeed, it was at the first workshop at Fermilab that the idea of colliding linac beams was first seriously considered and at which new phenomena such as beamsstrahlung were recognized. It was also at these workshops that a 20 TeV proton synchrotron-collider, i.e., 40 TeV c.m., was first discussed and proclaimed to be practical. Perhaps now the time has come to build a World Laboratory for a 50 TeV c.m. Proton Collider and for a 500 GeV Linear Electron Collider. Such a World Accelerator Laboratory would contribute to the mutual understanding and trust between people that will be required for us, if we are to survive our technology.

Recently, there has been discussion of the use of laser and accelerator beams as military countermeasures against bomb-carrying missiles. Although any alternative to the precarious balance of terror that now dominates our lives would appear to be attractive, upsetting that balance also has its terrifying aspects. An attractive possibility is that a truly international laboratory might be set up to investigate and develop such techniques as might contribute to the mutual security of nations during the period of deep cuts in nuclear weapons that must inevitably occur, or to the problems of nuclear proliferation. Perhaps the two kinds of laboratories—high energy and defensive—might be mutually supportive, might indeed be combined into one world laboratory. Such a laboratory might help to bring back the concept of a World Atomic Energy Commission that Oppenheimer, Lilienthal, and others put forward at the end of WWII.

7 Models, Hypotheses and Approximations

Rudolf Peierls

It is a well-known symptom of old age that students are unbelievingly young. A slightly less well-known symptom is that young physicists make such unreasonably bad physics. I am very conscious of this and of the likelihood that some of my critical remarks can be attributed entirely to my advanced age. Let me insist from the start that I am full of admiration for the recent developments in many parts of physics, but I am aware of some pitfalls.

In a paper I wrote a little while ago [1] I analyzed the various ways in which we use models in physics for different purposes and the troubles we can get into by mistaking one kind of model for another. I do not want to go into all these possibilities, but I want to stress that it makes a difference whether you are working on an approximation to a known theory, or on a hypothesis which might correspond to some as yet unknown basic equation, or on a model which will describe a situation qualitatively, without being right in quantitative detail.

Confusion can be caused by overestimating the range of validity of the rules which you are using, and I shall give some typical examples of that. But one also has to guard against the opposite mistake, of being too timid in learning something from an approach whose basis is not formally established. We should not forget what Pauli used to call "the

NEW DIRECTIONS IN PHYSICS
The Los Alamos 40th Anniversary Volume
ISBN 0-12-492155-8

law of conservation of sloppiness", which says that any first approximation is better than it has any business to be.

This kind of confusion is not new. In fact, to show that my criticism is not entirely due to my getting out of touch with modern developments, let me illustrate my points with some old examples. My first example is taken from solid-state physics, where in the 1920s Debye gave a very instructive formula for the specific heat of solids. He guessed a simple form for the frequency distribution of the lattice vibrations, which was designed to have the correct behaviour at low frequencies, and to contain the correct total number of modes. It therefore had to give the right answer for the specific heat at very low and very high temperatures, and it gave a qualitatively reasonable behaviour in between.

His formula was extremely successful in fitting experimental data, and for a number of years people tended to forget that they were dealing with a qualitative description only. The reason was that there were no precise measurements at very low temperatures, and, as the parameter of the theory, which depends on a suitable average of the sound velocity over directions, was not known, the formula in fact contained an adjustable parameter, which was used to fit the data at medium temperatures.

One of the first to make precise measurements at low temperatures was Francis Simon, and there discrepancies showed up. Figure 7.1 shows the experimental results [2], and a Debye curve fitted to the data. (I shall not go into how the Debye temperature for this fit was chosen.) The difference between experiment and theory was regarded as a deviation, and this naturally was in the shape of a hump, since it had to vanish at very high and very low temperatures. Now, a hump in the specific heat is the signature of a transformation, and Simon concluded that the substance in question underwent a change of state.

This was entirely due to a misunderstanding of the nature of the Debye theory. I quoted this example in a Lecture at Oxford, which happened to be the Cherwell-Simon Memorial Lecture, and I was in some doubt whether it was proper on that occasion to refer to a mistake made by Simon. I decided it was in order because it is just from the mistakes of great physicists that we can learn. To be fair, I searched the literature for a similar mistake by Cherwell, but failed to find one. I concluded that evidently Simon was the greater physicist of the two.

There are also old examples of the other kind. One case, in which an important idea was almost dropped, is well known: the suggestion of the electron spin by Uhlenbeck and Goudsmit. When they explained their idea to H.A. Lorentz, he liked it very much, but raised an objection. If the electron had a magnetic moment, it would also have to have a magnetic self-energy, and as the field of a dipole is more singular than

Model-making in physics   Figure 7.1

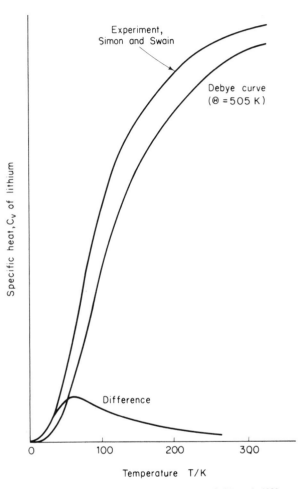

Source: Reprinted from *Contemporary Physics*, vol. 21, p. 1, 1980, with permission.

that of a point charge, one would have to make the electron radius rather larger than before, and such a large radius would not be compatible with spectroscopic data. The young inventors were horrified that they had committed such an elementary error, and wanted to withdraw their paper, but it was too late.

Another old example was pointed out to me by Steve Weinberg. In the 1930s quantum electrodynamics was in trouble because of the infinities that affected the calculation of any quantity beyond its leading order, and we therefore had the feeling that we could not meaningfully talk about any corrections to the leading term. This was sensible, although one might reasonably have speculated about the order of magnitude of the corrections. But in fact the horrors affecting any attempt to calculate higher terms led to the feeling that there were no such corrections. In particular when there were the first indications of a split between the $S_{\frac{1}{2}}$ and $P_{\frac{1}{2}}$ levels of hydrogen, these found very little credence. It was only when Lamb's experiments established the shift beyond any possible doubt that theoreticians realized that there was a correction, and the challenge posed by the problem of calculating this correction started off the development of modern quantum electrodynamics. Perhaps a little more faith in the powers of the theory might have gotten this development started much earlier.

A well-known example of not being deterred by apparently firm theoretical obstacles is provided by Ernest Lawrence. When he planned the cyclotron, he was thinking of the magnetic field as uniform. In that case it can be proved that the machine will not work, because the particle orbits will drift about. The project was saved by the unavoidable inhomogeneity of the field, which stabilizes the orbit. Lawrence certainly was not aware of this, and I suppose if it had been practically possible to make the field uniform he would have chosen to do so. Lawrence showed the same obstinacy in the face of theoretical objections when he designed the calutron. At the time it was well known that there was a maximum possible current that could be used in a mass spectrometer, before the space charge of the beam would spoil the separation. Lawrence ignored this and designed the machine for a current orders of magnitude greater than the theoretical limit—and it worked. It was understood only later that this was due to the residual gas in the "vacuum" chamber. The gas atoms are ionized by the beam; the electrons are trapped by the magnetic field, and the positive ions escape. In this way the space charge of the beam is neutralized.

I do not know what conclusion one should draw from these examples of Lawrence's attitude. I doubt whether one should conclude that it is right to ignore all theoretical arguments. Perhaps the right answer is that, just as one should keep an open mind for the possibility that we have overlooked an important argument, we should also keep an open mind for the errors in generally accepted negative conclusions.

Heisenberg was once asked what he would do when a theory disagreed with experiment. He answered that there were two possibilities: Either the theory in question was approximate or involved doubtful assumptions—then no conclusion was possible; or else it was a good theory—then the experiment was wrong. An admirable principle, but we may have some trouble deciding when a theory is good!

Turning from these ancient examples to more recent ones, it seems that the kind of confusion I mentioned has become more common. I believe there may be three reasons for this. One is the great variety of methods and approximations in use, which makes it more difficult to keep track of all their implications. Another reason may be that, of the younger theoreticians of today, fewer have had the experience of having to solve problems to which there exists a well-defined answer, requiring the use of approximate or numerical methods whose validity can be tested. A third reason is perhaps that we now have very little controversy in the literature. If you look at the journals of, say, 1920, you will find many short notes pointing out errors of reasoning in published papers. Many of these notes are by distinguished authors, including Planck and Einstein. This custom not only had the advantage of instilling some caution in authors who might publish papers too readily, but it also helped the reader to avoid following wrong arguments.

A particularly rich source of the kind of confusion I have in mind is the theory of transport phenomena, such as the electric or thermal conductivity of solids. Such problems can be tackled by the use of the so-called Boltzmann equation, which is a crude, but often effective, tool. The idea is to consider, for example, in the case of the electric conductivity, the number of electrons moving with a given wave vector at a particular time and determine the rate of change of this number due to the applied electric field and due to the collisions with phonons or impurities. In equilibrium the total rate of change must be zero. The validity of this type of equation is subject to certain conditions, but in many of the practical situations one can convince oneself that these are satisfied.

However, these methods are now regarded by many as old-fashioned. Indeed there are now much more sophisticated and elegant tools available. For example, there is the method of Green's functions. In a problem involving interacting particles, there exists an infinite set of coupled Green's functions, expressing respectively the physics of one, two, . . . , particles. These coupled equations are beautiful and elegant, but we cannot do anything with them as they stand. The accepted way of handling them is to use some "decoupling" procedure. This means stopping at the equation for a certain order, and replacing the Green functions of higher order occurring in it by products of lower-order functions, or by some similar device. At this point it is rather obscure what the approximation is and what kind of effects are neglected.

A similar device consists in writing down an infinite series representing the required quantity, which can be done very elegantly in terms of diagrams, and then summing a partial series of these diagrams. Sometimes the physical significance of such a subset of diagrams is clear (and then the answer usually could have been written down without using the series in the first place), but often it is not. A particular way of doing this is the so-called "random-phase approximation", which for some purposes can be shown to give at least qualitatively correct results, but by no means always.

A typical example of such problems is the thermal conductivity of non-conducting crystals, which I solved in principle in my 1929 Ph.D. thesis, [3], using of course the old-fashioned approach in terms of a Boltzmann equation, which, for all normal materials, turns out to be justified. The most interesting conclusion is that, in a pure crystal, the thermal conductivity rises exponentially at low temperatures. The reason is that the so-called "Umklapp" processes are essential for reaching a stationary state, since they involve a kind of Bragg reflection of the lattice waves in addition to the mutual scattering of different waves. In their absence, the sum of the wave vectors of all phonons is conserved, like the sum of momenta of molecules in a gas. In that case the collisions between phonons could not cancel the phonon drift caused by the temperature gradient. The Umklapp processes require the participation of very energetic phonons, which are rare at low temperatures.

More recently, many experts were dissatisfied with the old-fashioned treatment and felt it necessary to solve the problem by modern methods. A typical example is the work of J. Ranninger, [4], who succeeded in setting up a solution in terms of summing a cleverly chosen sub-series of diagrams. He found, to his satisfaction, that the conductivity tended to infinity at $T = 0$. However, it rose only as an inverse power of T, which was not right. It further turned out that his method, if applied to an elastic continuum instead of a crystal, when there are no Umklapp processes, would still give a finite conductivity. We had long and heated arguments before I convinced him that his answer could not be right, because it contradicted elementary conservation laws. He finally agreed, and found some other terms which could be included and which produced a more reasonable answer [5].

There are many other examples of papers which proceed by some formalism because it has been used before (and preferably one which has a name) but without considering whether it contains the essential ingredients for the problem in hand, and which prefer this "modern" approach to the old-fashioned one in which the nature of the approximations is transparent.

A somewhat different instance arose in the theory of the electric conductivity of metals. The old familiar formula for the conductivity is

$\sigma = ne^2\tau/m$, where n is the density of conduction electrons, m their effective mass, and τ a collision time to whose definition I shall return. There exists a very elegant formula due to Kubo, which is, in principle, an exact expression for the conductivity [6]. It is, however, rather abstract, and not easy to evaluate. In particular, if the collisions which limit the conductivity are in some sense weak, one would like to use an approximation in which they figure as a perturbation, and this is done in the old-fashioned Boltzmann approach. But you could not directly expand the Kubo formula in terms of the collision probability, because it is an expression for the conductivity, which, in the absence of collisions, is infinite. A direct expansion from infinity evidently does not make sense.

In 1959, G.V. Chester and A. Thellung set themselves the task of evaluating the Kubo formula for the case in which the electrons were scattered by random scattering centres (as in the case of very impure metals) using the elegant methods developed by L. van Hove. After some hard work, they came up with the above result for the conductivity. In their result, the collision time τ was the mean time between collisions, i.e., $1/\tau$ was the probability per unit time of any collision taking place. It was then pointed out that this result was correct only if the scatterers acted like hard spheres, i.e., if the differential cross section was independent of angle. In the more general situation, in which the cross section depended on the angle between the initial and final direction, one knows from the old-fashioned approach that $1/\tau$ should be the collision rate, times the average value of $1 - \cos\theta$, θ being the scattering angle. So they looked at their calculations again and found that there were some cross terms that they had not noticed before, and allowing for these, they reproduced the correct weight factor. Then I asked what would happen in an anisotropic medium, in which the scattering cross section depends both on the initial and the final direction. The old-fashioned Boltzmann equation in this case becomes an integral equation, that has no solution in closed form. The modern method therefore could be correct only if it produced the answer in a form related to the solution of this integral equation.

So these authors applied their method to the case of an anisotropic medium, and, being excellent physicists, they found indeed that the answer now involved the exponential of a certain operator, which could be evaluated only by reducing this operator to its diagonal form. The transformation doing this turned out to be the solution of the same integral equation as the Boltzmann approach [7].

This example shows that modern methods, used carefully, can produce correct answers. They do not always represent an easy route to the answer, and, in the present case, knowing the answer from the old-fashioned solution was of great help.

But let me also quote an example of the opposite from solid-state theory. In all work on electronic transport phenomena one always neglected the interaction between the electrons, except for some average effects through their space charge. We knew of course perfectly well that the Coulomb interaction was much stronger than many effects we did allow for, and if I am asked why we persisted in neglecting it, I do not have a clear answer. In part this was because everyone else was doing it— which is just the attitude I have criticized. In part it went back to the old classical theories, in which the electron-electron interaction was disregarded, because it conserved momentum and therefore did not contribute to the resistance. This was a misleading argument, because in the solid it is the wave vector rather than the momentum that is conserved, and this conservation law also holds for electron-phonon collisions, which do contribute to the resistance. Our procedure was justified later by Landau's Fermi-liquid theory. If we had refused to go on until we had understood the point, it would have held up progress.

Next I shall turn to nuclear physics. Here one of the examples of a model whose range of validity was misunderstood is Niels Bohr's original form of the liquid drop model. Bohr's idea of the state of a compound nucleus in which there is close coupling between all available degrees of freedom was of great importance in explaining the physics of nuclear collisions, particularly of neutron-nucleus collisions. But everybody, including Bohr, then thought that this meant there was always such strong coupling, so that the nucleus was like a little drop of water. This made the idea of a shell model appear quite ridiculous, as ridiculous as thinking that in a water drop you could, to any reasonable approximation, regard each molecule as pursuing its orbit independently of the others. Bohr believed this so firmly that when in 1937 he encountered Yamanouchi in Japan, who had some shell-model type of analysis of nuclear levels, he told him that this could not possibly make sense, causing Yamanouchi to abandon the approach.

It was only when Maria Mayer and others assembled the evidence for the existence of magic numbers, and later when she and Jensen and others succeeded in accounting for the values of these magic numbers, that the shell model became respectable. The difference between the low-lying states of nuclei, for which the shell model makes sense, and the highly excited states, such as those formed by a neutron entering a nucleus, for which it does not, was then seen to lie in the different level densities in the two cases. The nucleon-nucleon interaction is less than the typical level spacing near the ground state, but vastly greater than the level spacing of the highly excited system.

New trouble came when experiments showed that the internucleon potential contains a strong repulsive core. This meant that at close distances correlations between nucleons certainly could not be ignored, and this seemed to conflict with the validity of the shell model. This contradiction was resolved by methods initiated by Brueckner and extended by Bethe and others. These methods generalize the Hartree-Fock approximation, which treats each particle as moving in the average field of the others, by treating the interaction of any pair of particles as taking place in a suitable average field due to the others. The methods are ingenious, but not easy to apply quantitatively, particularly to finite nuclei.

The fashion has therefore arisen of starting papers on nuclear spectra with an incantation to the Brueckner-Bethe method, and stating that the interaction potentials used in the rest of the paper are to be understood as effective interactions in the Brueckner-Bethe sense. After this the authors proceed to do their calculations as if no problem existed. They conveniently ignore the fact that the effective potentials introduced in that method will depend on the nucleus, on its state, and on the location within the nucleus. The neglect of these features is usually hidden by the fact that the resulting many-body problem is too hard to solve without further simplifying assumptions, so that one does not expect a high accuracy in the results.

Another example of a pervasive fashion in nuclear physics is the use of the harmonic oscillator well. This has the advantage that the one-particle energies and wave functions can be written down in closed form, and for low-lying states in light nuclei it is a good approximation. Even for heavier nuclei, for which the shape of the potential may be rather different, it remains a good model for qualitative conclusions, and its simplicity allows often very instructive discussions. But one would like to see in each case attention paid to the question whether the special shape of the potential is likely to affect the kind of conclusion one is trying to draw.

There is a device of a similar nature, the so-called quadrupole-quadrupole interaction. If the interaction of two nucleons is replaced by such a term, which is not claimed to be an approximation to the real interaction, the secular problem arising in the shell model becomes soluble in closed form, and therefore this model is very instructive in making it possible to exhibit qualitative features which are typical of real nuclei. This model has therefore been extremely useful. But from time to time one sees papers in which this interaction is used for quantitative calculations, without considering how far it is appropriate for the purpose in hand.

The third example of this kind is the use of separable interactions, i.e., whose matrix elements factorize into a product of a factor depending on the variables of one state, and a factor depending on the other. To avoid misunderstandings I should perhaps make it clear that I am referring to separable potentials of first order, because the same name is sometimes used to denote interactions which can be expressed as sums of such products. A sufficiently large number of such terms can of course be made a good approximation to any real interaction, but it loses the simplicity of calculation allowed by the single product. Here again is a model which is very rich in instructive answers for many processes, but its use without appropriate reflection about its suitability for a particular purpose can mislead.

Finally let me quote an example from particle physics, though I feel less at home in this than in the older fields. I have in mind the "Veneziano model". The scattering amplitudes for elementary processes must satisfy certain conditions which, apart from the conservation laws for momentum, energy, etc., include the "crossing symmetry", an analytic connection between processes generated from each other by replacing the absorption of one particle by the emission of its antiparticle. It was not easy to see what form the amplitude should have to satisfy this requirement. Veneziano [8] was able to construct a function which did this, and this was certainly instructive.

He knew it could not be the right answer because it did not satisfy another important requirement, that of unitarity. Veneziano's work inspired a number of people to try extending his results so that they would retain the crossing symmetry while also satisfying unitarity. This was worth trying, although so far these attempts do not seem to have been successful.

But others took the Veneziano amplitudes as they stand to make predictions about various processes although they could not be the correct answer, and called this the "Veneziano model".

Let me add a comment on a situation in particle physics, which I found amusing. When the quark model became attractive because it could explain so many basic properties of the hadrons, one began to wonder why no free quarks had ever been observed. To account for the absence of such evidence the concept of confinement was introduced, without first being backed by a full formalism. Then Fairbanks [9] announced experiments which appeared to show the existence of free quarks. One might have expected particle physicists to welcome these results as confirming the quark model and making the assumption of confinement unnecessary. But instead the general reaction was that the experiments could not be right, because it was known that quarks were confined. This was a surprising reaction, but it is even more surprising that it seems to have been justified.

So what lessons can we learn from these and similar examples? Some seem to tell us to be careful and avoid using models and approximations lightheartedly. Others tell us to be bold and not to be intimidated by doubts. Well, there is no short-cut to the decision of what is right and what is wrong. Don't be carried away either by beautiful and elegant approximations or by seemingly fatal objections. Don't be afraid of living dangerously but make sure you know when you do.

References

[1] Peierls, R., (1980) *Contemporary Physics*, **21**, 1.

[2] Simon, F., and Swain, R.C., (1935) *Z. Phys. Chem.* **28**, 189.

[3] Peierls, R., (1929) *Ann. Physik* **3**, 1055. The Quantum Theory of Solids, Oxford 1955, Chapter 11.

[4] Ranninger, J., (1967) *Ann. of Physics* **45**, 452.

[5] Ranninger, J., (1968) *Ann. of Physics* **49**, 297.

[6] Kubo, R., (1957) *J. Phy. Soc. Japan* **12**, 570.

[7] Chester, G.V., and Thellung, A., (1959) *Proc. Phys. Soc.* **73**, 745.

[8] Veneziano, G., (1968) *Nuovo Cimento* **57A**, 190.

[9] La Rue, G.S., *et al.*, (1977) *Phys. Rev. Lett.* **38**, 1011.

8 Comments on Three Thermonuclear Paths for the Synthesis of Helium

Anthony Turkevich*

In the universe as we understand it today, helium is synthesized on a large scale out of lighter nuclei under two conditions: in the first few minutes during the expansion of the universe from its earliest high temperature state and in stars such as our sun. In addition, the small amounts of deuterium (and other thermodynamically unstable light nuclei) surviving from the nucleosynthetic period of the early universe can be concentrated by human activities and ignited to complete their conversion into helium. This has been accomplished under the conditions of fusion weapons and is the aim of the controlled thermonuclear energy programs.

All three scenarios, Big Bang, stellar interior, and fusion weapon or controlled fusion, involve thermonuclear reactions. It is of interest to compare the thermodynamic conditions and time scales of these three processes and to note that different fundamental interactions of nature play dominant roles in the three situations.

The first path to be examined is the generally accepted process by which the sun produces energy. This is considered to be the main process in many other stars as well. It is a likely process to have operated in the first stars, when negligible amounts of heavy element seeds were available for alternate paths.

* Supported in part by a grant from the US Department of Energy.

NEW DIRECTIONS IN PHYSICS
The Los Alamos 40th Anniversary Volume
ISBN 0-12-492155-8

In the sun the fueling mixture is about 75% protons and 25% helium (by weight), with only a few percent of heavier elements. The presently accepted series of reactions [1] occurring in the center of the sun is based on a first step originally suggested by Atkinson [2] and first calculated in detail by Bethe and Critchfield [3]:

$$P + P \rightarrow D + \beta^+ + \nu \tag{1}$$
$$D + P \rightarrow {}^3He + \gamma \tag{2}$$
$${}^3He + {}^3He \rightarrow {}^4He + 2P \tag{3}$$

(this last reaction has an alternative (~14%) path that need not concern us here). The rate-determining step for this chain is the first reaction. Even at the high central solar density (~156 g cm^{-3}) and temperature (~1.5 × 10^7 K) the reaction rate is only about a million reactions per gm per second, leading to a solar lifetime of billions of years. The rate constant for reaction (1) is determined by the probability for positron emission during the short time that the protons are close enough during a collision and therefore by the strength of the Fermi interaction. The importance of this first step can be illustrated by noting that the lifetime of a deuteron in the center of the sun, limited by the rate of reaction (2) is more than seventeen orders of magnitude shorter than that of the proton, limited by the rate of reaction (1). The collision frequencies and Gamow penetration factors entering into the estimation of these lifetimes are similar in the two cases.

The early universe as a locale for dynamic elemental synthesis was first postulated by Gamow [4] in the late 1940s. Detailed calculations on the course of nuclear synthesis under such conditions were carried out first by Fermi and Turkevich [5]. More than ten years later, the difficulty of explaining the observed ~25% universal abundance of ^{4}He by stellar processes [6] and the discovery of the 3 K background radiation [7] led to a re-examination of the nucleosynthetic scenario at the time of the early universe. Although numerous nuclear reactions involving the original protons and neutrons and the many intermediates are possible and contribute, the important reactions in this scenario during the nucleosynthetic stage are the following:

$$n \rightarrow H + \beta^- + \bar{\nu} \tag{4}$$
$$n + H \rightleftharpoons D + \gamma \tag{5}$$
$$D + D \rightarrow {}^3H + H \tag{6}$$
$$\rightarrow ({}^3He + n) \rightarrow {}^3H + H \tag{7}$$
$$D + {}^3H \rightarrow {}^4He + n \tag{8}$$

In the current Standard Model for this primordial nucleosynthesis, summarized, for example, by Schramm and Wagoner [8] the initial particles are an approximately 7/1 mixture of protons and neutrons determined by the conditions at any earlier, higher temperature, stage of the universe. The baryon density and temperature follow the time behavior given by

$$n_b = \frac{n_b^*}{t^{\frac{3}{2}}} \tag{9}$$

$$T_9 = \frac{13.8}{t^{\frac{1}{2}}} \tag{10}$$

where T_9 is the temperature in units of 10^9K, t is the time in seconds, and n_b^* is the baryon density in particles cm^{-3} at $t = 1$ sec and, although still uncertain, is of the order of 10^{22}.

Nucleosynthesis starts a few minutes after the initial universe singularity, when the temperature drops sufficiently (to $<10^9$ K) for enough deuterium to be present to allow the above reactions to proceed beyond the deuterium stage. During this period most of the density of the universe is in the form of electromagnetic radiation, which determines the temperature-time relation (10). The rate of helium production is determined by the concentration of deuterium [kept almost in the equilibrium amounts represented by (5) and by the rate of the deuterium destruction processes (6–8)]. Nucleosynthesis proceeds until most of the neutrons are consumed and the expansion of the universe, with resultant drops in concentration and temperature, quenches the reaction mixture, leaving unreacted traces of D, ^{3}H, and ^{3}He as well as very small amounts of nuclei heavier than the ^{4}He. The main production of ^{4}He occurs in a time period of about 30 sec. The gravitational, electromagnetic and strong interactions are all involved in determining this time scale.

To these two naturally occurring paths for helium synthesis was added the man-initiated thermonuclear fusion program in the 1950s. This involves the use of deuterium and other thermodynamically unstable light nuclei, some of which are considered to be unreacted residues of the Big Bang. If heated to a high enough temperature [9], these can react thermally to form the stable ^{4}He nucleus.

As an example, we consider the nuclear reactions that occur when deuterium at liquid densities (or higher) is heated to temperatures in excess of 10 keV:

$$D + D \rightarrow {}^3H + H \tag{6}$$

$$\rightarrow {}^3He + n \tag{7'}$$

$$D + {}^3H \rightarrow {}^4He + n \tag{8}$$

It is seen that these are very similar to the ones that occurred after the first steps in the early universe. All three of these reactions produce two heavy particles (as do the controlling steps in the thermonuclear reactions

Table 8.1. Characteristics of Thermonuclear Paths for Helium Synthesis

Locale	Temperature (K)	Density (g cm^{-3})	Time Scale (sec)	Ref.
Thermonuclear Fusion	$\sim 10^8$	$>1.4 \times 10^{-1}$	$<10^{-5}$	
Early Universe	$\sim 8 \times 10^8$	$\sim 0.6 \times 10^{-5}$	$\sim 3 \times 10^1$	[8]
Sun	0.16×10^8	1.6×10^2	$\sim 10^{17}$	[1]

when other fuels are used). The transition probabilities governing the rates (after Coulomb penetration) depend on the strong forces between hadrons. In the thermonuclear weapons regime, where the initial densities are relatively high, the time scale for the reactions is less than 10^{-5} sec. In the projected controlled thermonuclear fusion programs, where the densities are relatively low, this time is longer, but still less than 10^{-3} sec.

Some characteristics of these three thermonuclear paths for helium synthesis are summarized in Table 8.1. It is seen that under the conditions where these processes occur the temperatures differ by less than a factor of a hundred, and the densities vary by seven orders of magnitude. However, the time scales involved cover more than 22 orders of magnitude. This span reflects to a large extent the relative strengths of the Fermi interaction (important in the solar reactions) and the hadronic interaction (important in the thermonuclear fusion reactions). The early universe reaction time scale lies between these two extremes and involves the gravitational, electromagnetic, and hadronic interactions.

References

[1] Bahcall, J.N., Huebner, W.F., Lubow, S.H., Parker, P.D., and Ulrich, R.K., (1982) *Rev. Mod. Physics* 54, 767.

[2] Atkinson, R.d'E., (1936) *Astrophys. Jour.* 84, 73.

[3] Bethe, H.A., and Critchfield, C.L., (1938) *Phys. Rev.* 54, 248.

[4] Gamow, G., (1948) *Nature* 162, 168.

[5] Fermi, E., and Turkevich, A., (1948) unpublished calculations quoted in detail by R.A. Alpher and R.C. Herman, (1950) *Rev. Mod. Physics* 22, 153.

[6] Hoyle, F., and Taylor, R.J., (1964) *Nature* 203, 1108.

[7] Penzias, A.A., and Wilson, R.W., (1965) *Astrophys. Jour.* 142, 420.

[8] Schramm, D.N., and Wagoner, R.V., (1977) *Ann. Rev. Nucl. Sci.* 27, 37.

[9] E. Teller [*Science* 121, 267 (1955)] gives credit to E. Fermi for suggesting, in the early 1940s, that a nuclear fission explosion could heat deuterium to the ignition point.

9　… And the Sad Augurs Mock Their Own Presage*

E. Segrè

I have been asked to contribute a short essay on the future of Science
and of Physics in particular. It is a perilous enterprise and also, I am
afraid, one not too suited for a person in his late seventies. Experience
hardly compensates the inroads of age and the fresher outlook of youth.

　I could try to escape by some fairly safe generalizations. I could refer
to Newton, who compared himself to "a boy playing at the seashore
and diverting myself in now and then finding a smoother pebble or prettier
shell than ordinary, whilst the great Ocean of Truth lay all undiscovered
before me", or else to Faraday, who answered a politician enquiring
about the use of his discoveries: "I do not know at present, but I am
sure one day you will be able to tax them." If Newton and Faraday had
to look for escape hatches, what should a common man do? Even Rutherford, when he tried to be specific, could blunder badly. In 1937, barely
five years before the Chicago pile went critical, he wrote: "The outlook
for gaining useful energy from the atom by artificial processes of transformation does not look promising." Or perhaps the great man had all
his prophetic spirit concentrated in the word *useful* and he anticipated
weapons and unreasonable environmentalists (both not useful)? Fermi

---

* Wm. Shakespeare, Sonnet CVII

NEW DIRECTIONS IN PHYSICS
The Los Alamos 40th Anniversary Volume
ISBN 0-12-492155-8

in 1923, the year he got his doctorate, was more farsighted. He wrote, referring to the transformation of matter into energy: "One will rightly say that it does not seem possible, at least in the near future, that one will find a way of liberating such tremendous amounts of energy. We can only wish that it be so, because the release of such a tremendous amount of energy would have as its first effect the blowing to pieces of the physicist so unlucky as to find the way of accomplishing it."

In physics as in every other discipline, there are branches that are getting old and losing interest. It has always been so in the past, and it will continue to be so, with one caveat. Occasionally an old branch that seemed half dead may bloom again, for instance by the injection of new techniques that allow one to see details that had previously escaped. Atomic physics is an example and not the only one.

I believe that particle physics is still at the center of interest, at least intellectually. It has scored memorable triumphs in the past decades and revealed great surprises such as the nonconservation of parity. Although it has not generated as great a deepening of our understanding or an epistemological change as quantum mechanics, it has nevertheless penetrated two layers of the divine onion that seems to symbolically describe the world. As in all of physics, progress occurs through an interplay of theory and experiment. No other efficient method is known. However, in high energy physics, the increasing cost and complexity of experiment poses a real problem, not only a technical one, but one that transcends science. The reactions of society to such a poser, through government decisions, will be of paramount importance. It is a new situation because only recently the sums of money involved in "pure" scientific research have become significant on a national scale, although still very small compared to government budgets.

For the immediate future there will be a great amount of work to be done on the standard model and related subjects.

I have heard predictions, mainly by theoreticians, that until much larger energies are available, i.e., until one reaches the realm of grand unification, there will be a desert. Here experience helps me to disbelieve. I have heard similar predictions many times, even by illustrious masters, and they were always in error. I surmise that the "desert" will prove full of "oases".

In other fields of physics, one sees progress on very complicated problems. The computer has opened up new possibilities and has established a hybrid between theory and experiment that may give totally unexpected results. In the immediate future one sees a confluence of new mathematical theories and computing techniques applied to many body problems. Furthermore, nonlinear problems are revealing new features, and the world is full of nonlinear effects. Is this going to reveal new physics? Optimism is warranted. Some of the great minds of physics, from Maxwell to our contemporaries such as Feynman and Yang, have straddled statistical physics and field theory, the two main phyla of physics, and not in vain. The two phyla now even start to interact, and this might become an interesting nonlinearity.

We can expect further successes in the race to very low temperatures. The field has been prolific in new results, even practically important.

A little farther afield we have astrophysics. Experiment is here replaced to a large extent by observation, but in what a stupendous laboratory and under what extraordinary conditions! There is no hope of duplicating on Earth the ambient conditions we observe on many stars, thus we must be content with what we can see. It is likely that observation from space will give an increase and improvement to our powers of detection similar to what the telescope has over the naked eye and that the astrophysicists may repeat the words of Galileo on his observations "since they are infinitely stupendous, I am infinitely grateful to God who has deigned to make me the first observer of things so admirable and hidden to all past ages". Here also the necessary costs are high. How much will a modern society be willing to pay for something that neither kills us nor promises to heal our diseases? The irrational basis by which we allocate our means does not give any way of predicting.

Most important extensions of physics loom on the horizon. We cannot forget the triumphs of molecular biology nor the cultural background and training of several of its foremost practitioners, who started their career as physicists.

In the scientific enterprise, one finds ever deeper and longer-range connections. Certainly nobody could have predicted that x-rays would be fundamental to genetics twice, first for production of mutations and later for disentangling the molecular basis of the gene, nor that nuclear bombardments by the creation of radioactive indicators would pervade biology at all levels. Science is developing such recondite couplings that it reminds one of a living organism.

Physics has been my strong vocation since a very early age, but possibly I would not have been able to make it my profession if it had not been for the lucky circumstances prevailing at Rome in 1927. The choice of a scientific profession depends not only on the aptitudes of the person and on the problems at the center of attention, but also on external circumstances. This, at least, is the rule for common mortals. A Rutherford, a Faraday, or a Mendel may be irrepressible, provided they are given the slightest chance. There are also other important factors: A person willing and suited to work in one of the large collaborations of modern science could probably fit into a modern industrial job and might prefer it for economic reasons, the more so because the large collaborations deprive their members of several of the attractions of single handed experimental work. Biology, at least from a distance, seems still to offer opportunities for scientists who are not very sociable and adaptable to large collective enterprises, a trait common in many distinguished scientists of the past. On the other hand, the biological cast of mind is truly different from that of a physical scientist and certainly not interchangeable.

In conclusion I want to reemphasize the great speed and unpredictability of scientific and technical progress. The discovery of the atomic nucleus, Bohr's atom, and quantum mechanics are all younger than I. I was already an active physicist at the time of the discovery of the neutron. I saw some of the first computers at Los Alamos. I learned as a student some Mendelian genetics, but I remember the impact of the discovery of the double helix. None of these things could have been foreseen at the time I was born. If we contemplate technological achievements the wonder is even greater. Thus I expect that my grandchildren, if they ever read these pages, will shake their heads at my shortsightedness. Let us hope that this great race, however, will not end catastrophically. The great progress of science and technology have increased greatly the power of mankind for good and for evil. If there is a collective free will we should see to it that it will be used to choose our common good. Otherwise, there might be no future.

10 Experiments on Time Reversal Symmetry and Parity

Norman F. Ramsey

Introduction

Some symmetry tests are sensitive to both time reversal symmetry (T) and parity (P) while others are sensitive to P alone. The present report is concerned with two groups of experiments of the first kind and one of the second. These experiments are:

Tests of T and P
    Neutron electric dipole moment
    Proton electric dipole moment and P and T violating effects in TlF
Measurements Violating P and Not T
    Parity violating rotations of the neutron spin by the weak interaction

Neutron Electric Dipole Moment

A neutron electric dipole moment would occur if there were a slight bulge of positive charge distribution in one hemisphere of the neutron and of negative charge in the opposite hemisphere. It is easy to see that the existence of such an electric dipole moment would violate both T and P. Since the only means for specifying the orientation of a particle

NEW DIRECTIONS IN PHYSICS
The Los Alamos 40th Anniversary Volume
ISBN 0-12-492155-8

is by the orientation of its spin angular momentum, $\mathbf{J}$, the electric dipole moment μ_E must be given by

$$\mu_E = C\mathbf{J} \qquad (1)$$

Therefore, if time is reversed, $\mathbf{J}$ is reversed and hence μ_E is reversed. Consequently, the electric fields resulting from μ_E would be reversed while ordinary electrostatic fields would be unchanged, so the magnitude of the resultant electric field would be different in the time reversed system, and the physics in that system would be changed, contrary to the assumption of time reversal symmetry. Consequently, a search for an electric dipole moment of either the neutron or proton is a test of T. A similar analysis shows that the existence of an electric dipole moment is also incompatible with P. For both of these reasons it was originally argued [1] that the neutron could not have an electric dipole moment, but the author pointed out in two early papers [2,3] that both T and P symmetry were assumptions that should depend on experimental confirmation and that the neutron electric dipole moment provided a sensitive check. Subsequently, the work of Lee and Yang [4] and of Wu, Ambler, et al. [5] showed that P symmetry indeed was not valid for the weak interaction in beta decay. The later experiment of Fitch, Cronin, Christianson and Turlay [6] showed that there was a violation of CP symmetry in the decay of the long lived neutral kaon, K_L°, and hence there should be a violation of T symmetry if CPT were conserved, as usually assumed. So far, all manifestations of T violations have been exclusively in the K_L° and the possible discovery of a neutron electric dipole moment appears to be the most sensitive means of detecting a T violation in another particle. At present the low limit on the neutron electric dipole moment provides a severe restriction on many theories of particles and forces. For this reason, there is a continuing effort to reduce the limit on the neutron electric dipole moment as much as possible.

The reason that the neutron provides a more sensitive test than the proton is its zero charge. If a particle has an electric dipole moment its orientation dependent interaction energy is given by

$$W = -\mu_E \cdot \mathbf{E} \qquad (2)$$

Consequently, an electric field is required to observe the electric dipole moment of a particle. But a charged particle will be accelerated out of the apparatus by an electric field or, if the charged particle is the nucleus of a neutral atom, it will readjust its position in the atom until $\vec{E} = 0$. An exception to this statement can occur if the particle is acted on by another force. This, for example, is the case of, say, the odd proton in a Tl nucleus which is subject to both electric and nuclear forces. This exception provides the basis for the *TlF* experiment discussed below. However, even in this case the electric field is almost cancelled and the neutron appears to be the hadron for which the lowest electric dipole limit can be established.

All of the neutron experiments I am discussing depend on the use of totally reflecting neutron mirrors, so let me first discuss some of the properties of such mirrors. Since a wave is associated with the neutron there is an index of refraction n when a slow neutron passes through matter. The index of refraction is given by

$$n = \left[1 - \frac{\lambda^2 N a_{coh}}{\pi} \pm \frac{\mu_M B}{\frac{1}{2} M v^2} \right]^{\frac{1}{2}} \tag{3}$$

where λ is the neutron wave length, N the number of nuclei per cm^3, a_{coh} is the neutron coherent forward scattering length, μ_M the neutron magnetic moment, B the magnetic induction and $\frac{1}{2} M v^2$ the neutron kinetic energy. As in the case of fibre optics, total reflection occurs for a glancing angle θ greater than the critical glancing angle θ_c where, as for light

$$\cos\theta_c = n \tag{4}$$

as shown in Figure 10.1. A neutron conducting pipe can then be made

(a)

(b)   NEUTRON PIPES:

(c)   NEUTRON BOTTLE FOR $v_n < 6$ m/s:

6 m/s ↔ 2x10^{-7}e V ↔ 0.002K ↔ 670Å

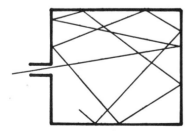

Figure 10.1. (a) Total reflection of neutrons. (b) Neutron pipes. (c) Neutron bottle.

to transmit all neutrons at a smaller glancing angle. The departure of n from unity increases with λ^2 so for very slow neutrons the neutron pipes are particularly effective.

Since the critical angle by Eqs. (3) and (4) depends on the orientation of the neutron spin in the magnetic inducting B, a short length of magnetized mirror provides a highly effective polarizer and analyzer.

Finally, at neutron velocities below 6 m/sec there is total reflection even at normal incidence and the neutrons can be stored in a bottle for 100 sec or longer.

For many years the most sensitive measurements of the neutron were succession of neutron beam magnetic resonance experiments that have been previously described [7,8]. The neutron electric dipole moment was measured by observing the shift in the neutron magnetic resonance frequency when the relative orientations of the static electric and magnetic fields were reversed. The most sensitive of these experiments was that of Dress, Miller, Pendlebury, Perrin and Ramsey [7]. It gave the results

$$\mu_E/e = (4.0 \pm 15) \times 10^{-25} \text{ cm} \qquad (5)$$
$$|\mu_E/e| < 30 \times 10^{-25} \text{ cm}$$

In other words, the neutron electric dipole moment, if it exists at all, is less than 30×10^{-25} cm. The smallness of this result can be appreciated by noting that for this limit the corresponding bulge by one unit of positive charge in one hemisphere of the neutron would correspond to only 0.01 cm if the neutron were expanded to the size of the Earth.

The above neutron beam electric dipole moment experiments depended upon the fact that neutrons at a velocity of, say 80 m/sec were totally reflected by many materials at glancing angles of 5°. When the velocity is less than 6 m/sec total reflection can be obtained even at normal incidence on many surfaces and it is possible to store neutrons in an enclosed bottle. For many years it was apparent [9,10] that electric dipole moment experiments with bottled neutrons would be particularly sensitive, but for a long time suitable sources for adequate quantities of ultra-cold neutrons were not available.

The first experiment with bottled neutrons to provide a useful limit to the neutron electric dipole moment was that of Altarev and his associates in Leningrad [11]. They used the apparatus shown in Figure 10.2 with the neutrons stored in a double-neutron bottle. The electric fields in the two halves of the double bottle are in opposite directions to provide a first-order cancellation of magnetic-field fluctuations. The authors used a new form of the separated oscillatory field method called the adiabatic method (Ezhov et al., 1976) of separated oscillatory fields. In this method the static magnetic fields in the oscillating field regions are inhomogeneous

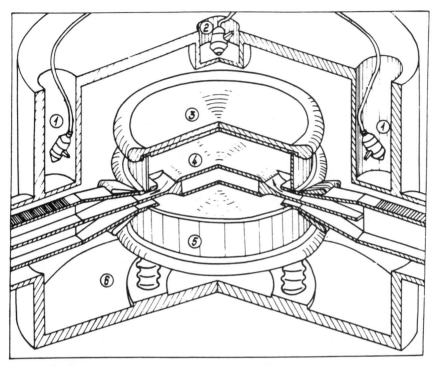

Figure 10.2. General view of the Leningrad double-bottle neutron magnetic resonance spectrometer: 1 indicates a magnetometer for control of magnetic field, 2 a magnetometer for field stabilization, 3 the top electrode, 4 the central electrode, 5 the chamber wall made of beryllium-oxide-coated fused quarts, and 6 the bottom electrode. The initial and final oscillatory field coils are shown on the neutron pipes at the left- and right-handed sides of the figure. The spectrometer is surrounded by three layers of magnetic shielding.

in such a way that the neutrons adiabatically reach the resonance condition in the first field and depart from it adiabatically in the second region. As in the neutron beam experiments, the electric dipole moment is inferred from the shift in resonance frequency when the relative orientations of the static electric and magnetic fields are reversed.

Typical resonances obtained in the Leningrad experiment are shown in Figure 10.3. The published results of this experiment [11] are

$$\mu_E/e = (2.3 \pm 2.3) \times 10^{-25} \text{ cm} \qquad (6)$$
$$|\mu_E/e| < 6 \times 10^{-25} \text{ cm}$$

A preliminary value of $(-2 \pm 1) \times 10^{-25}$ cm has more recently been given [11].

At the Institut Laue-Langevin (ILL) at Grenoble, a Harvard-Sussex-

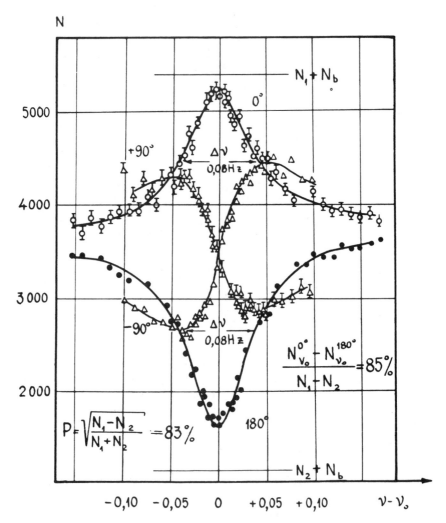

Figure 10.3. Typical resonance with Leningrad spectrometer.

Rutherford-ILL collaboration is also using bottled ultra-cold neutrons at less than 6 m/sec to measure the neutron electric dipole moment. These neutrons are led by a neutron-conducting pipe into the apparatus shown in Figure 10.4. The neutrons are stored in a cylinder approximately 25 cm in diameter and 10 cm high with the top plates being metallic—beryllium or copper—and the sides of the cylinder being insulators of beryllia or quartz. After the neutron bottle is filled, a shutter is closed storing the neutrons for 30–100 s. The oscillatory field is applied as initial

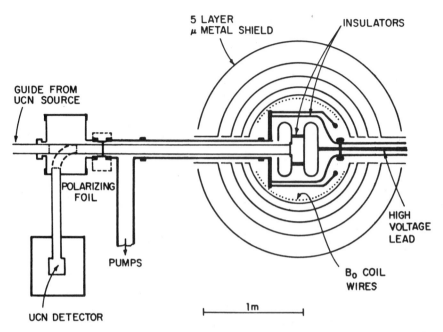

Figure 10.4. Schematic diagram of apparatus for measuring the neutron electric-dipole moment with bottled neutrons at the ILL.

and final coherent pulses, usually with a $\pi/2$ relative phase shift. The resonance is observed in a fashion similar to the neutron beam experiment: the neutrons are polarized on passage through the indicated polarizing foil, analyzed during their return passage through the foil, and counted at the indicated ultra-cold neutron (UCN) detector. The observations are at the steepest point of the resonance curve. The change in beam intensity correlated with the application of an electric field is then examined to set limit to the neutron electric-dipole moment. A typical resonance curve obtained with the new ILL apparatus is shown in Figure 10.5. The published result of the ILL experiment [23] is

$$\mu_E \neq e \neq (-68 \pm 2.9) \times 10^{-25} \, cm$$

The beam intensity has subsequently been increased by more than a factor of 100, reducing the statistical error in the experiment to 0.7×10^{-25} cm. Possible systematic errors, however, are still being studied, and no new value has been announced.

The use of stored ultra-cold neutrons offers two particularly important advantages. The resonance curve with stored neutrons in Figure 10.5 is 3500 times narrower than that with the neutron beam apparatus. Fur-

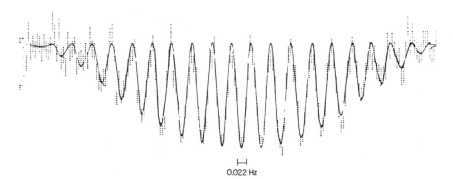

H
0.022 Hz

Figure 10.5. Neutron magnetic resonance obtained experimentally with bottled neutrons. Two phase-coherent but successive oscillatory fields were applied and the change in the neutron intensity was observed as the oscillatory frequency was varied. The observed resonance width of 0.022 Hz is to be contrasted to 80 Hz of Figure 10.2.

thermore, a large fraction of running time in the beam experiment must be devoted to eliminating the $\mathbf{E} \times \mathbf{v}/c$ effect. Since it is the average value of $\mathbf{v}$ that is important, this effect is drastically diminished when the neutrons enter and leave by the same exit hole with an 80-sec storage time instead of passing through the apparatus once at a velocity of 90 m/sec. As a result of the reduced effective magnetic field from $\mathbf{E} \times \mathbf{v}/c$, it is also possible to use a much weaker static magnetic field with an accompanying reduction in the stability requirement for the current that provides the static magnetic field.

Although the new experiments with bottled neutrons have the above marked advantages, it must be recognized that they still have many difficulties. The limit has by now been pushed to such a low value that care must be taken to avoid all possible systematic effects. Although some of these are intrinsically reduced in an experiment with bottled neutrons, other serious problems remain. For example, problems due to stray magnetic field (especially when associated with reversals of the electric field) and to magnetic-field changes resulting from electrical sparks can be just as serious in absolute terms and more serious in relative terms with bottled neutrons. These problems have already caused much difficulty in the beam version of the experiment and should be even more formidable in the bottled-neutron experiments, which seek a much lower limit for the neutron electric-dipole moment. Golub and Pendlebury [13] have proposed the use of cold liquid ^{4}He in a neutron bottle to accumulate ultra-cold neutrons in one bottle while the resonance is observed in another. This procedure should provide much greater neutron densities in the bottle where the resonances are studied. However, as the sensitivity limit is pushed to successively lower values, the problem of magnetic field fluctuations, especially any that are coupled jointly to the electric and magnetic field reversals, will probably determine the limit to D that can be reliably established. How low this limit can be will be determined by experience during the next few years. A proposal by Ramsey [14] to use optically pumped ^{3}He or a similar gas as a magnetometer should contribute markedly to lowering the limit on μ. The ^{3}He would be optically pumped to align the nuclear spins and then introduced into the neutron bottle at low pressure along with the neutrons. Then oscillatory fields would be applied simultaneously at the neutron and ^{3}He resonance frequencies. The ^{3}He would be subsequently pumped to a high magnetic field region where its polarization would be measured in an NMR spectrometer. In this fashion the ^{3}He would measure the magnetic field in the same region of space and at the same time as the neutron measurements, except for a small correction for the effect of gravity on the ultra-cold neutrons. Heckel and others have suggested that ^{199}Hg could be more sensitively detected with lasers.

Almost all theories before 1964 predicted zero for the electric-dipole moment of all particles since theories were usually required to be symmetric under P prior to 1957 and under T prior to 1964. As discussed in the introduction, a theory that is symmetric under either P or T must necessarily lead to a zero value for the electric dipole moment of any particle. Subsequent to the experiments performed in 1957 [4,5] showing the existence of P symmetry violations in the weak interactions, and in 1964 [6] showing violations of CP symmetry in the decay of K_L°, almost all theories have predicted nonzero values for particle electric-dipole moments. Although the experiments of Christenson et al. [6] imply a violation of CP symmetry and the experiments of Schubert et al. and Casella imply a violation of T symmetry, both are limited to the neutral kaon system. Failures of CP or T symmetry are yet to be found in any other system. As a result, it is not yet established as to which of the fundamental interactions is associated with the CP and T violations. It is even possible that these violations are not in any of the established interactions but are instead in a new superweak interaction. For this reason it is not surprising that the different theories predict widely different values for electric-dipole, depending on the model assumed.

Kobayashi and Maskawa (1973) showed that a quantum chromodynamics gauge theory based on only four quarks and a single Higgs scalar multiplet cannot account for even the known CP and T violations in K_L°. Therefore, most current theories attribute the T violation either to a six (or more) quark theory or to the exchange of multiple Higgs bosons.

The predictions of various theories are presented graphically in Figure

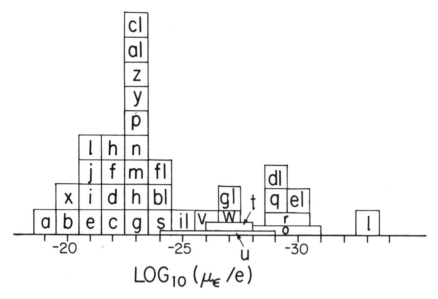

Figure 10.6. Theoretical predictions of the neutron electric-dipole moment. Each lettered block corresponds to a different theory with the references to the different theories given in reference [8].

10.6. Each theoretical prediction is represented by a block of equal area so the height of the block is correspondingly diminished if the prediction spans several decades. From Figure 10.6 it is apparent that it is highly desirable to lower the experimental limit on the neutron electric-dipole moment to distinguish between the different theories. In particular, of the two categories of theories most favored at present, the ones that attribute the *CP* violation to the exchange of Higgs bosons predict values of $D \sim 10^{-24}$ cm, whereas those that attribute it to additional quarks give $D \sim 10^{-28}$. The experiments now in progress should soon distinguish between these two categories of theories.

Electric Dipole Moment of Proton and *T* Violating Effects in *TlF*

As discussed in the previous section, the average electric field in the proton is zero if it is acted on by no other forces and is not accelerated. However, the odd proton, p, in the *Tl* nucleus of the *TlF* molecule can be subject to a net electric field on p since the electric force on it can be balanced by the nuclear force of the other nucleons, provided the electric field is sufficiently inhomogeneous. The internal molecular electric field is strongly inhomogeneous and can be oriented by the application of a strong external electric field. This is shown schematically for a molecule with $\dot{m}_J = -1$ in Figure 10.7, where $\langle \mathbf{B}_p, \uparrow \rangle$ is the average magnetic induction for proton p when the strong electric field points upward and $\langle \mathbf{E}_p, \uparrow \rangle$ is the same thing for the average electric field. From this figure and from assumed time reversal symmetry

$$\langle \mathbf{B}_p, \uparrow \rangle = \langle \mathbf{B}_p, \downarrow \rangle \tag{7}$$

and

$$\langle \mathbf{E}_p, \uparrow \rangle = -\langle \mathbf{E}_p, \downarrow \rangle \tag{8}$$

Therefore, if the Tl nucleus has both a magnetic moment and a small electric dipole moment contributed dominantly by proton p, the precession frequency will be increased in the presence of a strong electric field in one direction relative to the axis of the molecular rotation and decreased in the opposite direction.

This change in the resonance frequency can be detected by the molecular beam magnetic resonance method. However, for a high beam intensity it is best to use focusing electric quadrupoles and these do not give a significantly different deflection for molecules in states (J, m_J, m_{Tl}, m_F) $= (1, -1, \frac{1}{2}, -\frac{1}{2})$ and $(1, -1, -\frac{1}{2}, -\frac{1}{2})$. This difficulty can be overcome

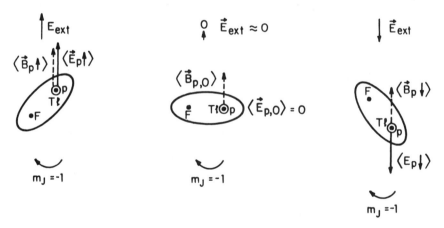

Figure 10.7. Schematic representation of a TlF molecule in rotational state $J = 1$ and $m_j = -1$ in an external electric field strongly upward in the figure on the left, approximately zero in the middle and strongly downward to the right. In all cases the molecule rotates in a counterclockwise direction about a vertical axis. $\langle \mathbf{B}_p, \uparrow \rangle$ is the average magnetic induction for proton p when the strong electric field points upward and $\langle \mathbf{E}_p, \uparrow \rangle$ is the same thing for the average electric field. By time reversal symmetry $\langle \mathbf{B}_p, \uparrow \rangle = \langle \mathbf{B}_p, \uparrow \rangle$ and $\langle \mathbf{E}_p, \uparrow \rangle = -\langle \mathbf{E}_p, \downarrow \rangle$.

by deflecting the molecules in the state $(1, 0, -\frac{1}{2}, -\frac{1}{2})$ and by inducing the transitions $(1, 0, -\frac{1}{2}, -\frac{1}{2}) \leftrightarrow (1, -1, -\frac{1}{2}, -\frac{1}{2})$ immediately before and after the main oscillatory field region. A reorientation of the nuclear spin in the main oscillatory field region can then be detected by the impossibility of a subsequent transition and by the consequent failure of focussing. With such an apparatus the shift in frequency on reversal of the electric field can be detected.

Although an experimentally observed shift in the resonance would probably be interpreted as evidence for a neutron electric dipole moment, it could also be due to a failure of time reversal symmetry in the atom, in which case Eq. 7 would be invalid. However, such a failure of time reversal symmetry in atomic forces would be of comparable interest to a neutron electric dipole moment.

The above method was first used by E.A. Hinds and P.G.H. Sandars [15]. Although the first experiment benefitted from a very long beam, it suffered from extremely weak intensity. As a result the experiment has been repeated on a shorter apparatus with a much more intense beam by Wilkening, Larson, and Ramsey [16] with the results for the proton

$$\mu_E/e = (1.3 \pm 2.0) \times 10^{-21} \text{ cm} \qquad (9)$$

The experimental error was limited by some systematic errors, possibly due to shifts in the effective beam position in the apparatus. The authors believe that with an improved apparatus design to eliminate the systematic errors the final experimental error could be reduced by a factor of ten or more.

Parity Non-Conserving Rotations of the Neutron Spin

In 1964 F. Curtis Michel [17] pointed out that in principle a neutron spin polarized perpendicular to the neutron's velocity would precess about the velocity vector in a parity non-conserving manner due to the weak force when the neutron passed through matter; a decade later L.S. Stodolsky [18] independently reached the same theoretical conclusion.

If ϕ_{PNC} is the angle of rotation after passing through a sample of length l, the rotation should be given [17–19] by

$$\phi_{PNC}/l = (n_+ - n_-)/\lambda = -2\pi\lambda N_a(a_+ - a_-)$$
$$= \sqrt{2}(\hbar/Mc)^2 N_a G M^2 W$$

where

$$W = Z C_{ne} + |Z C_{np} + (A - Z) C_{nn}|\eta$$

and

$$C_{ne} = 1.25 \, (2 \sin^2\theta_w - \tfrac{1}{2})$$

where n_+ is the index of refraction for the neutron wave with the spin in the direction of motion, λ the neutron wave length, N_a the number of atoms/cm³, a_+ the forward scattering length for the neutron spin in the direction of motion, M the neutron mass, G the Fermi constant $= 1.02 \times 10^{-5}/M^2$. W is the effective weak current charge where C_{ne} is the effective charge of the neutron-electron interaction, C_{np} is the same for the n–p interaction, C_{nn} is the same thing for the neutron-neutron weak interaction, and θ_w is the Weinberg angle.

For many years, the papers of Michel [17] and Stodolsky [18] were mostly ignored because of the anticipated smallness of the effect—about 1.4×10^{-8} radian per cm of bismuth traversed. Most experimentalists considered this rotation to be far too small to be observable. Nevertheless, when Stodolsky first called the author's attention to these calculations, it became apparent that the sensitivity for detecting the rotations of the neutron spin in the various neutron electric dipole moment experiments was greater than that required for detecting the parity violating rotations predicted by Michel and Stodolsky.

A proposal was made to the Institut-Laue-Langevin to look for such a parity non-conserving rotation in a $^{209}_{83}$Bi rod of 1 m length where the predicted rotation was 1.4×10^{-6} radian. Bismuth had the attractive feature that its low cross section for slow neutrons permitted the use of long samples and hence provided high sensitivity. M. Forte of Ispra, Italy, later suggested that due to a low lying 62 eV resonance, a much larger effect should be seen with $^{124}_{50}$Sn. Since the effect being looked for had never been seen before in any substance, it was clearly desirable to look for it first where it should be large. Furthermore, the effect proposed by Forte with ^{124}Sn should have been big enough to be seen with a weaker neutron beam and consequently could be looked for earlier.

As a result, Forte joined the group that made the original proposal and a search for the effect in $^{124}_{50}$Sn was started. By the time the experiment was actually to begin, Blayne Heckel, the Harvard graduate student doing most of the work, raised serious doubts as to whether there should be enchancement in ^{124}Sn; however, there were sufficient doubts both about the theory and about the criticism of the theory that it seemed best to proceed with the experiment in any case.

A schematic diagram of the apparatus is shown in Figure 10.8. Neutrons are polarized and analyzed by curved stacked magnetized foils. The neutron is left in a particular orientation after passing through the first current sheet and passes through the Sn sample where the spin is rotated

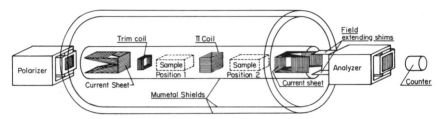

Figure 10.8. Schematic diagram of the apparatus used to measure the parity violating rotation of the neutron spin due to the weak interaction.

by the weak interaction. The neutron passes through the second current sheet into a coil where the orientation of the magnetic field is perpendicular to the field in the first sheet. The neutron then passes to the analyzer. Any spin rotation then changes the number of neutrons transmitted by the analyzer. That the rotation is due to Sn can be seen by the use of the π coil which rotates the spin π radians about the coil axis and hence converts an increase in intensity to a decrease and *vice versa* if the Sn is in position 1. On the other hand, if the sample is in position 2 there is no change in intensity when the π coil is turned off or on.

Runs were planned both with a ^{124}Sn sample and with an ordinary Sn sample as a control. The results of the experiment by Forte, Heckel, Ramsey, Green, Greene, Pendlebury, Sumner, Miller, and Dress [19] showed no observable rotation of the neutron spin in the ^{124}Sn sample but, surprisingly, did show a significant rotation in the ordinary Sn sample being used as a control. Although the observed rotation was only a five standard deviation effect, it was convincing and an abstract giving this result was published [19].

After this unexpected result was obtained, the group decided the effect was possibly attributable to the 7.7% of ^{117}Sn in the ordinary Sn sample, since the neutron capture γ rays in the isotope also show a surprising large parity anomaly. Therefore, a run was made with ^{117}Sn and a very large effect was seen. Subsequently, parity non-conserving rotations of the neutron spin have been seen in other materials as well.

If ϕ_{PNC}/l represents the parity non-conserving spin rotation per unit length of sample traversed, if ϕ_{PNC} is taken as positive for right handed rotations about the linear momentum vector, and if an atomic symbol without an assigned atomic number represents a natural sample of the element, the results obtained so far are

$$\phi_{PNC}/l\,(\text{Sn}) = -(3.19 \pm 0.40) \times 10^{-6}\text{rad/cm}$$
$$\phi_{PNC}/l\,(^{124}\text{Sn}) = -(0.48 \pm 1.49) \times 10^{-6}\text{rad/cm}$$
$$\phi_{PNC}/l\,(^{117}\text{Sn}) = -(37.0 \pm 2.5) \times 10^{-6}\text{rad/cm}$$

$$\phi_{PNC}/l \, (PB) = +(2.24 \pm 0.33) \times 10^{-6}rad/cm$$
$$\phi_{PNC}/l \, (La) = -(219 \pm 29) \times 10^{-6}rad/cm$$

At present, these experiments are temporarily interrupted while other experiments occupy the neutron beam. It is hoped that these experiments can later be resumed with emphasis on lighter elements and elements which have no nearby nuclear resonance to obscure the neutron-electron interaction by the resonant enhancement of the nuclear interaction.

References

[1] Rutherford, E., (1920) *Proc. Roy. Soc.* **A97**, 374.

[2] Purcell, E.M., and Ramsey, N.F., (1950) *Phys. Rev.* **78**, 807.

[3] Ramsey, N.F., (1958) *Phys. Rev.* **109**, 225.

[4] Lee, T.D., and Yang, C.N., (1957) *Phys. Rev.* **105**, 1671.

[5] Wu, C.S., Ambler, E., Hayward, R.W., Hoppes, D.D., and Hudson, R.P., (1957) *Phys. Rev.* **105**, 1413.

[6] Christenson, J.H., Cronin, J.W., Fitch, V.L., and Turley, R., (1964) *Phys. Rev. Lett.* **13**, 138.

[7] Dress, W.B., Miller, P.D., Pendlebury, J.M., Perrin, P., and Ramsey, N.F., (1977) *Phys. Rev.* **D15**, 9.

[8] Ramsey, N.F., (1978) *Phys. Rev.* **43**, 400, and (1982) *Ann. Rev. Nucl. Part. Sci.* **32**, 229.

[9] Ramsey, N.F., (1957) *Rev. Sci. Instr.* **28**, 57; (1958) *Phys. Rev. Lett.* **1**, 232; (1969) *Proposal* to ILL.

[10] Shapiro, F.L., (1968) *Usp. Fiz. Nauk.* **95**, 145.

[11] Altarev, I.S., Barisov, Yu.V., Brandin, A.B., Egerov, A.I., Exhov, V.R., Ivanov, S.N., Lobashov, V.M., Nazarenko, V.A., Porsev, G.D., Ryabov, V.L., Serebrov, A.P., and Taldaev, R.R., (1978) Leningrad Nucl. Phys. Inst. *Preprint* **430**, 1; (1980) *Nucl. Phys. A.* **341**, 269; (1981) *Phys. Lett.* **102B**, 13; (1984) *Jour. de Phys.* **45**, C3, 11.

[12] Harvard-Sussex-Rutherford-ILL. (1974). This collaboration includes C. Baker, J. Byrne, R. Golub, K. Green, B. Heckel, A. Kilvington, W. Mampe, J. Morse, J.M. Pendlebury, N.F. Ramsey, K. Smith, and T. Sumner. The equipment is described in *A Proposal to Search for the Electric Dipole Moment of the Neutron Using Bottled Neutrons.* Grenoble, France: Inst. Laue-Langevin (1974) and Oxford: Rutherford Lab (1975).

[13] Golub, R., and Pendelbury, J.M., (1975) *Phys. Lett.* **53A**, 133 and (1977) *Phys. Lett.* **62A**, 337.

[14] Ramsey, N.F., (1980) *Bull. Am. Phys. Soc.* **25**, 9.

[15] Hinds, E.A., and Sandars, P.G.H., (1980) *Phys. Rev.* **A21**, 471 & 480.

[16] Larson, D.A., (1981) Ph.D. thesis, Harvard University.

[17] Curtis Michel, F., (1964) *Phys. Rev.* **133B**, 329.

[18] Stodolsky, L.S., (1974) *Phys. Lett.* **50B**, 352.

[19] Forte, M., Heckel, B., Ramsey, N.F., Green, K., Greene, G., Pendlebury, J.M., Sumner, T., Miller, P.D., and Dress, W.B., (1980) *Bull. Am. Phys. Soc.* **25**, 526.

[20] Forte, M., private communication.

[21] Heckel, B., private communication.

[22] Ramsey, N.F., (1984) *Acta Physica Hungarica* **55**, 117.

[23] Pendlebury, J.M., Smith, K.F., Golub, R., Byrne, J., McComb, T.J.L., Summer, T.J., Burnett, S.M., Taylor, A.R., Heckel, B., Ramsey, N.F., Green, K., Morse, J., Kilvington, A.I., Baker, C.A., Clark, S.A., Mampe, W., Ageron, P., and Miranda, P.C., (1984) *Phys. Lett.* **136B**, 327 and (1986) *NBS Special Publ.* **711**, 25.

11   On the Course of Our Magnetic Fusion Energy Enterprise

D.W. Kerst

After the mid-1940s, the idea of controlled release of nuclear fusion energy began to attract the attention of physicists. The attempts to create proper physical conditions for fusion then became a continuing scientific enterprise of a very unusual character. It was only gradually realized that the problem was much more difficult than originally imagined, with the result that successive generations of physicists are involved. A physics development usually does not follow a course such as that taken by controlled thermonuclear research with a rush for experimental results without fairly well understanding the basic processes. Early thoughts on controlling fusion reactions came up at Los Alamos in Edward Teller's "wild idea seminar", which met periodically, with one of the topics being magnetic confinement. At one of these seminars where Jim Tuck, Robert Wilson, John Manley, E. Fermi, E. Teller and three or four others were present, Fermi created gloom about holding the ionized reactants in a toroidal magnetic field by pointing out the difficulty of confining the magnetic field lines themselves in the presence of extraneous fields—not to mention the plasma to be held in by these field lines. It was later that Spitzer's solution to this problem with the twisted stellarator field transform launched the Princeton effort. In this period at Livermore, Post's magnetic mirror studies began, and with Tuck's vigorous efforts

NEW DIRECTIONS IN PHYSICS
The Los Alamos 40th Anniversary Volume
ISBN 0-12-492155-8

using gaseous discharges at Los Alamos the work became known as the Sherwood project.

The early plasmas would not assemble themselves properly by simply closing the switch. It took quite a while to uncover some of the difficulties. At the first Sherwood information meeting in Denver in 1951 Teller could explain with a hand-waving argument that pinch discharges should be unstable to kinking. From then onward much more attention was paid to theoretical treatment of the stability problem.

This period from 1945 to 1958 was characterized by heroic attempts to create what could honestly be called thermonuclear neutrons with hydrogen isotopes. Our atomic energy commissioners could easily understand the value of this. But their expectations turned out to be too high to be reached by the time of the second "Atoms for Peace" conference in 1958. This conference at Geneva initiated an important change because, through the efforts of H.J. Bhabha of India, secrecy had been removed from controlled fusion work, and it finally became possible to learn what was being done in other countries through the many formal and informal discussions, along with exhibits and models of experimental equipment. The revelation of the scale and extent of Russian experimental work and their good theoretical work made a great impression.

In these early years decisive experimental results had been difficult to achieve. Interpretations of experiments and comparisons with theoretical expectations were helpful exercises, but often did not require belief. It was very difficult to connect theory and observation. This was to change during the 1960s.

Nevertheless, these early years produced a wealth of valuable experimental experience, for example, the ability developed to handle heroic electronics, high voltages, fast pulses and switching, heavy discharges, and trials with special magnetic configurations. There were linear and toroidal and theta pinch discharges, stellarators, mirrors, cusps, and plasma guns as sources. There was much good spectroscopy, but detailed measurements by internal field probing for use with Maxwell's equations did not start until approximately 1957. In 1957 James Tuck remarked that everything had been thought of by then. That could have been discouraging to the newcomer, but events in CTR moved swiftly enough to overwhelm any doubts.

Nevertheless, by 1962 it was generally believed that it was necessary to be more deliberate with trials of magnetic configurations and to pause to plan and to build a better experimental plasma physics base for the difficult work to come.

It was then time for wider participation in experimental work, for example, by universities. For decades there had been good experimental work on gaseous discharges and their plasmas, but it was not yet clear that acceptable experimental Ph.D. thesis work on confinement of hot plasmas could be carried out. The situation was different for theoretical work. The theory had a continuous development of its own, with an abundance of problems requiring scholarly attention, and the 1958 Geneva conference had shown that other countries—particularly England and Russia—also had produced excellent work.

But there is a characteristic of the enterprise which suggests some truth in Tuck's off-hand comment. For example, the very early discouraging toroidal pinch discharges, revived by adding a toroidal field, declined again in favor by 1960; they have now reappeared as reverse field pinches twenty years later with better understanding and with some very favorable properties. Some of the last toroidal discharge tests in the US in 1961 were run with the externally applied toroidal field much higher than the discharge's self field ($q \approx 1$ to 3) instead of the usual relatively low toroidal field ($q \ll 1$). Immediately afterward the Russian Tokamak operating in this new range, $q < 1$, became especially interesting and has since become one of the most favored configurations—it is a reincarnation of the earlier toroidal discharges. Likewise the mirror configuration has undergone declines in favor and subsequent revivals when additional possibilities for it became clear. The mirror is also one of the present day promising configurations. The stellarator, conceived and tried very early, has had a similar history. After much excellent experimentation, stellarator work stopped in the US, to be supplanted by Tokamaks. It did not perish in Europe, but it has now been revived in the US, and its attractive properties are now more clearly appreciated as it is again becoming an important part of the US program. Cusps, multipoles, Yin-Yangs, and baseball configurations are incarnations with different names of an early common topology. In several cases different names have been given to equivalent configurations like calked stuffed cusp and spherator. Almost cyclical variations in favor have occurred for most of these schemes.

Much of the present interest in compact tori for small reactors draws on experience with previous configurations and even on the bodily translation of plasmas as exhibited in the 1958 conference.

This resurgence of earlier configurations or ideas, which at one time seem dead-ended, gives us a proper interpretation of Tuck's comment about everything having been considered. The understanding of these schemes had to pass through successive stages of refinement. Our patrons and the newer members of this multigeneneration enterprise can take account of this characteristic of CTR in the attempts to get on with a best choice solution. Clear and decisive judgements cannot always be made at every stage of this enterprise. The nature of the problem so far appears to be such that many embodiments of magnetic confinement and of plasma manipulation have remarkable survival characteristics. Thus, in the future, these variations of configuration may all contribute components to a best solution, and there is the possibility that if one configuration can be made to work for a reactor, others might be made to work also—an optimistic but not unreasonable point of view.

Since the earliest reactor design requirement studies by R.F. Post in 1952, there were brief repetitions of engineering studies at intervals of several years. Generally the experimental plasmas were so far from having reactor parameters that little attention was paid to the ultimate engineering requirements. However, from 1972 up to the present, attention to reactor problems has been continuous and increasing in parallel with the developing body of experimental experience. The engineering realities brought to light have influenced decisively the direction and choices for experimental work.

Again we find a universally clear prize or goal—now of plasma ignition or break-even, propelling the program with high priority toward the goal—but this time with a more firm but nevertheless a still incomplete physics basis.

An onward impetus for swift action also came indirectly from the country's advances in the fission energy program. As David Rose prodded vigorously and influentially in the early 1970s: if a confinement configuration was not immediately chosen with a move into engineering development, then the country will have gone so far with fission reactor and breeder commitments that fusion may have lost its chance. However, now we see that the subsequent pace of investment in fission as a source of energy has not yet foreclosed choices for fusion.

For this engineering design of confinement systems great skill in using computers for simulation and fundamental plasma calculations has developed, and currently good progress is being made. However, we are not yet in a position to design some important aspects of a fusion reactor from the physical fundamentals. Experimental trial by construction and simulation models are still necessary. Now the need for discovering and understanding the physical processes acting is especially important before we have a clear view of the limitations and the possibilities for magnetically confined fusion reactions.

The components for producing reactor grade plasmas seem to be at hand—if not fully tried and tested. Sophisticated non-destructive diagnostics are well along in their development, several plasma heating methods are available, and there are also interesting untried heating schemes. There will no doubt be an extensive period of making things work with a parallel effort to make this energy source economical. Toward this latter goal the current guidance from the ultimate users, the electric utilities, is having an important influence. If this interest can be sustained throughout the trials ahead, the ultimate result should be acceptable in the energy market that the utilities serve.

12 Early Days in the Lawrence Laboratory (1931–1940)

Edwin M. McMillan

When I refer to the early days, I include only the period up to the end of 1940. By that time many people in the United States had become deeply concerned over the war in Europe, some had left the Laboratory for war work, and soon the Laboratory itself became involved in war work. One major peacetime project was started in 1940, the 184-inch cyclotron, but it did not get back to its original purpose until after the war. The hill above the Big C was chosen for the site, and by the end of 1940 the magnet foundation was completed and the bottom yoke was in place. This started the first expansion of the Laboratory off the campus, a stage in growth belonging to a later period.

The Radiation Laboratory was the personal creation of Ernest Lawrence. It was his idea. He got the financial support, he pulled together the equipment and drew the people, and of course he supplied the key idea, the cyclotron. Many other people helped in essential ways. I could name President Sproul of the University, Leonard Fuller of the Federal Telegraph Company and the University, who arranged the gift of the large magnet for the 27- and 37-inch cyclotrons, Frederick Cottrell and Howard Poillon of the Research Corporation, and Francis Garvan of the Chemical Foundation who looked with favor on Ernest's requests for grants, Raymond Birge who became Chairman of the Physics Department in 1932, Don

NEW DIRECTIONS IN PHYSICS
The Los Alamos 40th Anniversary Volume
ISBN 0-12-492155-8

Cooksey from Yale, Stan Livingston, and many others, but it was Lawrence's laboratory.

Those of us who were there in the early days remember that Ernest was always "the boss". He could be very rough on people if he felt they were not giving their utmost efforts, but he made up for this by his generosity in giving credit and in sharing ideas. I never met Rutherford, but I have been told that he had the same kind of character, with an important difference: Rutherford favored the individual researcher working with simple apparatus, while Lawrence believed in efforts so large that teamwork was necessary. In the very beginning there was a penalty for this. The drive for greater energy and beam current was so frantic that people hardly had time to think, and some important discoveries were missed and some mistakes were made, but this phase soon passed. On the whole, I think Lawrence was right; the rapid development of the cyclotron was more important to nuclear science than the question of who made which discovery.

The Laboratory was started in 1931, and when I came to Berkeley near the end of 1932 it was in full swing. There was not only the 27-inch cyclotron giving protons of around 2 MeV but also the Sloan x-ray tube, on which great hopes were placed for cancer treatment, and a couple of linear accelerators of the Wideröe type, which were built and operated by Dave Sloan, Wes Coates, and Bernard Kinsey. The Sloan x-ray tube was used clinically for several years, but the linear accelerator concept fell by the wayside, waiting to be revived by new ideas coming from wartime radar developments. It was certainly a busy place day and night, especially when Ernest was there, which was most of the time.

I started my research in Le Conte Hall on a molecular beam problem, but dropped that when the result I was seeking was obtained elsewhere and entered the exciting world of the Radiation Laboratory in the Spring of 1934. Stan Livingston, the cyclotron expert, and Telesio Lucci, a retired Commander in the Italian Navy who was a beloved general helper and factotum, gave me sage counsel on how to comport myself, as my previous experience had been in working alone, and I needed to learn the art of teamwork. This was not easy since no one was routinely coordinating the various tasks needed to keep the cyclotron going, and there were the twin dangers of neglecting what one should do and getting in the way trying to do something that someone else should do.

Robert Oppenheimer was the chief theoretical advisor for the Laboratory, and he suggested that I study the gamma rays produced by proton and deuteron bombardment of light elements. This turned out to be an important experiment, because I found a 5.5 MeV gamma ray from fluorine bombarded with protons, with which I could check Bethe and

Heitler's new theory of gamma ray absorption by pair production. The chief line of research then going on was the study of nuclear reactions by observing the protons and alpha particles which are emitted. These were detected by a thin ionization chamber connected to a linear amplifier, a device not suited for observing gamma rays.

Geiger counters were considered unreliable. Stan Livingston tells, in a paper presented in Texas in 1967, what happened on February 24, 1934, when the Laboratory learned of the Joliot-Curie discovery of artificial radioactivity. They were using a Geiger point counter, a device that now seems as exotic as the coherer, to count alpha particles. It was not the familiar cylindrical Geiger-Müller counter. The cyclotron oscillator and the counter circuit were turned on and off by the two poles of a double-pole knife switch, for convenience in timing. Within half an hour the switching arrangement was changed so that the counter could be turned on while the cyclotron was off, the counter voltage was raised so that it would count beta particles, the internal target wheel was rotated to bring a carbon target into the beam, and the activity of nitrogen-13 was there, produced by a different reaction than that used by its discoverers. The failure to see this first was a blow to the Laboratory, and there was a natural reaction against all Geiger counters.

So, the first thing I did was to go to Pasadena to learn from Charlie Lauritsen himself how to make quartz-fiber electroscopes. I had my first Lauritsen electroscope, which was mounted in a lead-walled chamber for detecting gamma rays, inside the laboratory for only a few days when Malcolm Henderson came to me in the middle of May with the news of the discovery of neutron-induced radioactivity in Rome, and he wanted to use it to look at some of those activities. It took only a short while to make a new chamber out of a tin can with a thin aluminum window, and the tin-can version of the Lauritsen electroscope became a valuable instrument for observing beta rays. Jack Livingood made one like it, which he used in a monumental survey, with Seaborg and others, of activities produced in many elements by deuteron bombardment, which resulted in a rate of discovery of radioisotopes that was comparable to that at Rome following Fermi's first neutron-induced activity. Some of those they found became very important in medical and other applications, like iodine-131, iron-59, cobalt-60, and technetium-99m.

There was a great surge of activity in the field of artificial radioactivity. The names involved are too many to list completely. Stan Livingston and I found a radioactive form of oxygen, and Lawrence found sodium-24, which created a sensation because very strong samples could be made. Lawrence once had the cyclotron crews working around the clock to make a whole curie for a demonstration. That was a tough job. Jackson

Laslett found sodium-22, the longest-lived artificial activity known at the time, but soon to be surpassed. Martin Kamen and Sam Ruben found carbon-14, probably the most important radioisotope of all, and so on. Also, new types of activities were discovered. Van Voorhis found that copper-64 could decay by emitting either negative electrons or positrons, the first known example of that kind of branching in radioactive decay, and Luie Alvarez found the first case of decay by orbital electron capture, now a well-known process.

Among the new activities were some that had atomic numbers differing from that of any known element and were therefore new elements. The first of these was found by Emilio Segrè and Carlo Perrier in 1937. They worked in Palermo with a piece of molybdenum that had been on the leading edge of the deflector plate in the cyclotron, where it got a lot of bombardment, and which Segrè had taken back with him after a visit in the summer of 1936. In it they found element 43, which they named technetium, after the Greek word for "artificial", as it was the first artificially produced element. Next was element 85, called astatine, from the Greek word for "unstable", found by Segrè, Dale Corson, and Ken MacKenzie in 1940.

A little later in the same year Phil Abelson, who had been a graduate student with Lawrence, came back to Berkeley for a short visit and supplied the missing link in the chemical identification of an activity induced in uranium by neutron bombardment, which had been puzzling me for some time. This was, as I had expected, the first transuranium element. I named it neptunium after the planet Neptune, just as uranium had been named after the planet Uranus.

After Phil left I continued the work, trying also the deuteron bombardment of uranium, which produced a different isotope of neptunium than the neutron bombardment, and found alpha particle activity in the neptunium samples, which suggested the presence of the next transuranium element. I did some chemical separations showing that the alpha activity did not belong to uranium or neptunium, but did not complete this investigation because I was persuaded by Ernest to go to the Massachusetts Institute of Technology for a few weeks to help set up a new laboratory for developing microwave radar (the term was coined later). As a cover, the new laboratory was called the Radiation Laboratory. Two Rad Labs! That was sometimes a source of confusion. I left by train on November 11, 1940.

On November 28 Glenn Seaborg wrote me that Art Wahl had been making some strong neptunium samples, and said: "If you are too busy to carry on the work alone we would be glad to collaborate with you." In my reply, on December 8, I said: "It looks as if I shall not be back in Berkeley for some time, and it would please me very much if you could continue the work on 93 and 94." (The "some time" stretched out to five years before I came back to stay. I never did believe Ernest's estimate of a few weeks.) On March 8, 1941, Glenn wrote to me describing the final chemical proof that the alpha activity belonged to the next element up the periodic table, plutonium. In this correspondence we did not use the names for the new elements, which were not yet official, but referred to 93 and 94, and the March letter was marked "Confidential". Secrecy was creeping into nuclear research.

Luie Alvarez came from Chicago in 1936, with a lot of clever ideas. He was the originator of the method of getting what is effectively a beam of very slow neutrons by pulsing the cyclotron and gating the detector so that it is only sensitive at some chosen time after the pulse of neutrons has been emitted. With Ken Pitzer, he used this method in an investigation of neutron scattering by the two kinds of molecular hydrogen, ortho and para hydrogen, and with Felix Bloch of Stanford he made the first measurement of the magnetic moment of the neutron. One of the questions of that time was the relative stability of the nuclei hydrogen-3 (called tritium) and helium-3, both of which had been observed by Mark Oliphant at the Cavendish Laboratory as products of the bombardment of deuterons by deuterons.

Alvarez and Bob Cornog first showed that helium-3 is the stable one by detecting it in atmospheric helium, using the cyclotron as a mass spectrometer; then, knowing that hydrogen-3 must emit beta-particles, they looked for activity in deuterium gas bombarded by deuterons and found it, establishing tritium as a radioactive isotope.

Gilbert Lewis of the UC Berkeley Chemistry Department played a very important role in the Laboratory's history. As soon as the discovery of deuterium was announced, he set up equipment to make heavy water by electrolysis and furnished a sample of heavy water to the Laboratory, and in March, 1933, the first beam of deuterons was produced by the cyclotron. From then on a major part of the work was with deuterons, which are much more prolific in producing nuclear reactions than are protons or alpha particles. Lewis, like many associated with the Laboratory, was a colorful character. He liked to tell how he fed some of his first heavy water to a fly, and it rolled over on its back and winked at him. One day at lunch at the Faculty Club he heard some professors in the department of education arguing about whether children should be taught to add a column of figures from the top down or the bottom up. Lewis said, "The way I do it is, first I add them down, and then I add them up, then I take the average."

I could go on and on. There were many visitors to the Laboratory who stayed and worked there for considerable periods of time, like Jim Cork from Michigan, Jerry Kruger from Illinois, Lorenzo Emo, a Count from Italy, Harold Walke and Don Hurst from the Rutherford Laboratory, Wolfgang Gentner from Germany, Maurice Nahmias from France, Sten von Friesen from Sweden, Ryokichi Sagane from Japan, and Basanti Nag from India. The working visitors were very important to the Laboratory. They not only contributed to the research program, but they also carried back the cyclotron art to their own institutions. Lawrence actively promoted this diffusion of knowledge, Don Cooksey wrote "cookbooks" of cyclotron lore which were mailed to interested institutions, and many Laboratory people went out to help design and build cyclotron laboratories. Milton White went to Princeton, Henry Newson to Chicago, Hugh Paxton to Joliot's laboratory in Paris, Jackson Laslett to Copenhagen, and Reg Richardson and Bob Thornton to Michigan.

Many of the physicists took part in the running and maintenance of the cyclotron. There were regular crews assigned to this task. I remember being on the owl crew for a while, which did not bother me as a single man with rather nocturnal habits but was hard on some others. If anything went wrong we had to pull the cyclotron apart and try to fix it. The greatest problems were vacuum leaks and the burnout of filaments in the ion source inside the cyclotron tank, and in the demountable oscillator tubes built by Dave Sloan. When the ion-source filament went out, the vacuum tank of the cyclotron had to be rolled out of the magnet gap, then the wax joint between the lid and tank broken and the lid removed, and the filament replaced; then it all had to be put back together again. Physicists did more than just operate the machine. For example, Art Snell and Ken MacKenzie built oscillators, Bob Wilson made the first theoretical study of orbit stability, and I designed the control system for the 60-inch cyclotron. This was in 1938, and a new concrete building, the Crocker Laboratory, was under construction to house the new larger cyclotron. The Laboratory was starting to expand.

Bill Brobeck came in 1937 as the first professional engineer hired by the Laboratory. That created a real revolution. No more waxed joints that leaked, no more equipment that fell apart in the middle of an important experiment. Or, at least, less than before. The "string and sealing wax" school of physics still has a nostalgic appeal to some old-timers like myself, but it is not suited to large efforts where many people are depending on the reliability of apparatus. Win Salisbury and Bill Baker, both electronic geniuses, took over the designing and building of oscillators and other electronic equipment. Charlie Litton came for a while and taught us many techniques in radio frequency engineering. He had a small company in Redwood City that he later sold to some entrepreneurs from Texas who used it as the nucleus for the giant conglomerate called Litton Industries. Charlie retired to Grass Valley where he spent the rest of his life happily working on various inventions.

Interest in biomedical applications started very early. Ernest's brother, John, is a physician, and Ernest always had an attraction to the field of medicine. I have already mentioned the Sloan x-ray tube, which went into medical use in 1934. The next year John came to Berkeley for the summer and made the first observations of the effects of neutron rays on a living organism, finding the effects greater than those of other forms of radiation and, therefore, very interesting, and in 1936 he came to stay. Paul Aebersold became the chief physicist for the biomedical group, making the arrangements for irradiation and measuring the dosage. The first cancer patient was treated in September, 1938, with sufficiently encouraging results that the Crocker Laboratory was dedicated to medical research, although the physicists and chemists got to use it too. There were working visitors in the biomedical field: Frank Exner from New York, Isidor Lampe from Utah, Raymond Zirkle from Pennsylvania, Al Marshak, Lowell Erf, John Larkin, and many others. Dr. Joseph Hamilton had a separate group studying the distribution of radioisotopes administered to animals and humans. To the smells of hot oil from the cyclotron were added those of animal colonies. As Laslett said in his "Cyclotron Alphabet", "M stands for mice whose smell makes us moan".

We went through the WPA period. It was during the great depression, and the WPA was a scheme by which unemployed people were hired by the Government and assigned to governmental bodies or other institutions to perform useful work. I have a 1934 letter from Lawrence to the University office handling this program, requesting, for a period of one month, "(1) One physicist, with Ph.D. and several years' subsequent research experience; (2) one carpenter; (3) one machinist, with several years' experience in general shop work." Some of those who came were real characters. I remember particularly Murray Rosenthal who was an amateur magician, a Swedish draftsman named Hallgren who was so profane that we tried to keep him away from Don Cooksey who objected to his language, and a man who had been with the telephone company, very distinguished looking, who liked to go around checking the strength of soldered wire joints by pulling at the wires with a buttonhook. Some, who were only temporarily down on their luck, stayed on and became valuable members of the Laboratory staff. Some idea of the financial scale of that time is given by the cost estimate made by Wally Reynolds in 1931 for the installation of the 80-ton magnet. This includes moving the magnet from San Francisco and setting it in place, four transformers, a 50 kilowatt motor-generator set, a 10-ton crane, concrete piers, labor, engineering, and contingencies, all for $5300!

It is hard to convey the atmosphere of that time. The world was in a deep depression. There was a general strike in San Francisco in 1934. Some people on the campus took sides, and friendships were broken over this. There was a lot of leftist agitation which later had dire consequences for many scientists. There was not much money around; for seven months, between the end of my fellowship and my appointment to the faculty as an instructor, I was a research associate without pay. But we all managed somehow, and the Laboratory kept going. Lawrence was the driving force, and the spirits inside the Laboratory were kept high by the excitement of discovery. There was very little organization; Lawrence was the boss, and that seemed to be enough. What a change has taken place since then! The eager youth has grown into an adult with the increased powers and problems that come with maturity.

Illustrations

Ernest Lawrence, on September 19, 1930, just after he had given the first scientific paper on the cyclotron at a meeting of the National Academy of Sciences on the Berkeley campus. He is holding a glass, brass, and wax apparatus with which he and Neils Edlefsen had obtained evidence of ion resonances in a magnetic field, encouraging Ernest to go on with the development of the cyclotron idea.

Here are Stan Livingston and Ernest Lawrence standing beside the big magnet in the shop of the Pelton Water Wheel Company in San Francisco. The magnet had been built by the Federal Telegraph Company of Palo Alto for use as part of a Poulsen-arc radio transmitter ordered by the Chinese government, but it was never delivered, and Leonard Fuller, a vice president of the company, arranged for it to be given to the University. The core and poles of the magnet had to be changed before it could be used as a cyclotron magnet, and that is being done here, in late 1931.

Stan Livingston made the first cyclotron that worked. He found a beam of 80,000 electron volt hydrogen molecular ions on January 2, 1931, in a 4-inch cyclotron. Then he made an 11-inch cyclotron, with which, in 1932, Milt White confirmed the lithium disintegration results of Cockcroft and Walton. This work was done in Le Conte Hall, but the big magnet needed a larger place to house it. Stan was one of the discoverers (or inventors) of strong focusing, without which most of high-energy physics could not have been done.

The Old Radiation Laboratory. It had been a civil engineering testing laboratory and was scheduled to be torn down, but Ernest persuaded President Sproul to let him have it for his experiments. This occurred on August 26, 1931, in President Sproul's office. At that time Ernest had the promise of financial support and a formal offer of the magnet, so if one wants to choose a day for a birthday this could be it. Early in 1932 the name "Radiation Laboratory" was painted on the doors. The magnet was installed in January, 1932, and the 27-inch cyclotron first operated in June of that year. Six years later the magnet poles were enlarged and the 37-inch cyclotron was installed. In the crew record for November 10, 1937, I found the following poem by Martin Kamen:

The cyclotron is a noble beast
It runs the best when you expect it least
Of all the pleasures known to man
The greatest is a good tight can

(by which he meant the vacuum tank).

In this building there was a large room for the cyclotron and its controls, an open court for transformers and switchgear, a machine shop, and some office space. Whenever there was trouble with the commutator of the generator that supplied the magnet current, I was called in to fix it. I was considered an expert at soldering with a torch

This is another view, taken in 1959. The demolition is proceeding toward you in the last view, and not much is left. I am standing there, sadly viewing the end of an era. Later, Crocker had to go too; the Chemistry Department needed space for more buildings.

---

in those days. Once Franz Kurie, when starting the motor generator, threw in the breakers in the wrong order and turned out the lights in all of Berkeley.

That building was the scene of frustration and elation—human as well as scientific drama. Many anecdotes have been told about happenings there, like the times that Ernest fired Bill Baker, and on an-other occasion Bob Wilson, only to recant and take them back again. But on the whole, relations were remarkably harmonious considering the many different temperaments of the people.

After the war, the first tests of the synchrocyclotron principle were done here, and in it Melvin Calvin did his pioneer work on the carbon cycle in photosynthesis.

The 27-inch cyclotron in 1932. The vacuum chamber sits between the poles of the magnet, covered with wax. The stovepipe going up in front has a wire strung down the middle that carries the collected beam current to a galvanometer on the control table out of the picture at the left. Sticking out in front is the linear amplifier built by Malcolm Henderson, which was used to count protons and alpha particles. The magnet windings were cooled by oil in the big circular tanks, and there was oil all over everything. One time Luie Alvarez neglected to close a valve after turning off the oil circulating pump, and a whole tank of oil ran over and went through the cracks in the floor into the basement.

Ernest at the other side of the cyclotron, also in 1932. Behind Ernest is the oscillator that supplied high-frequency power to the cyclotron. It used a commercial vacuum tube, but these were expensive, and for a while home-made tubes designed by Dave Sloan were used. Ernest is recognized as one of the world's great experimental physicists, but he was not particularly adept with his hands and contributed his share in the breakage of apparatus. When some delicate task was to be done, he would turn to someone else and say: "Here, you do this." His ideas and enthusiasm were the important things.

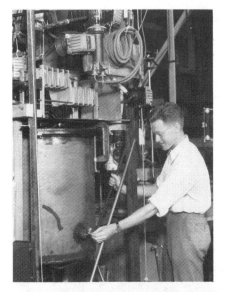

Dave Sloan with his x-ray tube, which was essentially a Tesla coil in a vacuum tank. Dave was very important to the Laboratory. He could build anything and was full of ingenious ideas. He built large oil-diffusion pumps when such items were not obtainable commercially, and made demountable oscillator tubes in which the filaments could be changed by taking apart a wax joint. One time he tried to make a diffusion pump using bismuth vapor. This did not work very well, but it was an interesting idea.

The machine shop in the old Radiation Laboratory. Without shops a laboratory could not operate. We used our own shop, the Le Conte shop, and large jobs were sent out to commercial shops. On the left is George Krause and on the right Eric Lehmann, working on a cyclotron tank, or at least looking as if they were contemplating working on it. Sitting in front are Don Cooksey, who made the shops one of his primary concerns, and Jack Livingood, the great hunter of radioisotopes.

This shows Art Snell, Franz Kurie, and Bernard Kinsey, who were, I think, in the Strawberry Canyon pool when this picture was taken. Art came from Montreal in 1934, later went to Chicago, and is now in Oak Ridge. He was famous as the poet laureate of the Laboratory; he would make limericks for all occasions. When Lawrence was awarded the Nobel Prize in 1939, he sent a wire that said: "Congratulations. Your career is beginning to show promise." He built an oscillator and discovered radioactive argon, among other things.

Franz Kurie from Yale seems to be giving a Tarzan yell, but he was actually a very gentle person. He introduced the cloud chamber tech- nique into the Laboratory. He made measurements of the energy distribution of beta rays and invented a method for presenting the data that made it easy to determine the upper limit of the energy. This is now known as the "Kurie plot" and has been widely used. In an investigation of the disintegration of nitrogen by neutrons, he found some unusual tracks which could be interpreted as being due to the capture of slow neutrons and the emission of protons, resulting in the formation of carbon-14. This served as a clue in finding the best method of making carbon-14 which, as you might guess from what I have said, is the capture of slow neutrons by nitrogen. I had a bottle of ammonium nitrate sitting

near the cyclotron target for a long time, hoping eventually to separate out carbon and see if it was active, but this got knocked over and broken, and I never put another one back. As I remember, people thought it was a nuisance and were afraid it might explode. There had been some large explosions involving ammonium nitrate. When carbon-14 was eventually identified as carbon bombarded with deuterons by Kamen and Ruben, they tried neutrons on nitrogen and never went back to the carbon bombardments, in which the yields were smaller and the active carbon was diluted by all the ordinary carbon. Franz later was the director of the US Navy Radio and Sound Laboratory in San Diego.

Bernard Kinsey was a Commonwealth Fellow from England, who built a linear accelerator for lithium ions. There are many stories about Bernard. He had a high temper and a very complicated and colorful form of swearing, really a high art. There was another Commonwealth Fellow at the University named Brown, who was probably the laziest man I ever knew. I don't think he ever did anything, but I saw him around the Faculty Club, where I was living at the time. He obviously was not in the Laboratory; Ernest would have thrown him out.

The Crocker Laboratory. The Old Radiation Laboratory is off to the right across an alley, and the 60-inch cyclotron resided in the high bay at the rear. This was called the medical cyclotron. It went into operation in 1939, giving deuterons of about 8 MeV. Under the supervision of Dr. Joseph Hamilton, it was used extensively for making radio isotopes for medical and tracer uses.

The 60-inch cyclotron, with Don Cooksey and Ken Green. It is much neater looking than the earlier cyclotrons. Bill Brobeck had had his influence. The structure projecting at the right is a pair of tanks that held the dee stems, which formed a resonant system. The oscillators were on the balcony at the right. The coil of heavy cable at the top carried high voltage to the deflector plate from the rectifier built by Ed Lofgren. The reason for the coil is that high voltage cables usually fail at the ends and are very hard to splice. The coil gave plenty of slack for making repairs.

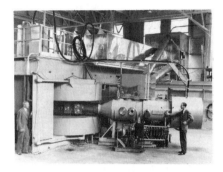

Looking through the window into the 60-inch control room. Bill Brobeck is there, Bob Wilson smoking his pipe, Ernest Lawrence, and a couple of other characters. The temporary setup that mars the neatness of the control table was a breadboard model of an automatic magnet current regulator that was being tested.

Ernest, Dale Corson, Winfield Salisbury, and Luie Alvarez, in 1939. Corson participated in the discovery of astatine, and later became the President of Cornell University. Salisbury has had a distinguished career in industry and the academic world since leaving the Laboratory; he made very valuable contributions to radar countermeasures during the war. Luie, of course, went on to win the Nobel Prize in physics.

This shows John Lawrence in 1936, with rows of mouse cages in the background, a proper setting for a biomedical researcher.

Again mouse cages, with mice, but a later date (1939) and a different person, Dr. Joseph Hamilton. Joe had a setup in Crocker where he worked with radioisotopes in medical and biological studies. Joe's table at the Faculty Club was noted for the interesting conversations on many subjects at lunch time.

All was not hard work; we had fun too. The Di Biasi parties were famous yearly affairs, at an Italian restaurant in Albany (Albany, California, that is). Paul Aebersold had an irrepressible sense of humor and was always the master of ceremonies. This party in 1939 was in celebration of the 60-inch cyclotron, and Paul was presenting a cake in the shape of a cyclotron, with the words: "8 billion volts or bust". That was supposed to be a wild exaggeration, but the Bevatron had not been invented yet. Lawrence and von Friesen are in the foreground.

Lorenzo Emo Capodilista, the Count from Italy, came to the Laboratory in 1935 and stayed several years. He did not use the last name, which means "head of the list" and is a name of great antiquity in Italy.

Charlie Litton in 1936, working with one of the glass lathes that he made.

Maurice Nahmias from Joliot and Curie's laboratory in France, with the vacuum chamber of the 37-inch cyclotron in 1937.

Henry Newson came from Chicago in 1934 with a Ph.D. in chemistry and turned into a physicist, doing some very ingenious experiments using the recoil of artificially produced radioactive nuclei. This picture was taken in 1938.

Ernest and Molly Lawrence, with Eric and Margaret, on the steps of Crocker Lab in 1939.

Ernest writing the script for a movie about his Nobel Prize in 1939.

Distinguished visitors in 1940: Lee de Forest and Diego Rivera.

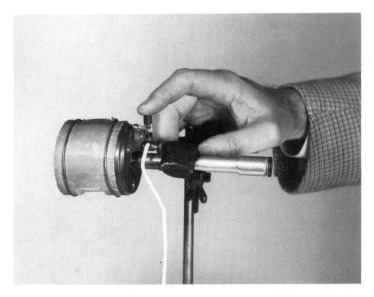

One of the original Lauritsen electroscopes that I made in 1934. (Now in the Smithsonian Institution in Washington.)

Me, at a press conference in Crocker Laboratory on June 8, 1940, at the time of the announcement of the discovery of neptunium.

I found this in the archives, and couldn't resist putting it in to end the show. I call it "On the Beach". Somewhere in the Sacramento River delta John Lawrence, Paul Aebersold, and I are enjoying the sun with some girls.

9

13 Nuclear Charge Distribution in Fission

*Arthur C. Wahl**

Introduction

The distribution of nuclear charge in fission is related to a number of other properties of nuclear-fission reactions, and these properties are reviewed briefly below.

It is well known for the more common low-energy fission reactions that division into two unequal masses, asymmetric mass division, is more probable than near symmetric mass division by about three orders-of-magnitude [1–3]. It is also known that the position of the heavy-mass peak remains essentially unchanged with increasing mass number of the fissioning nucleus; it is the light peak that shifts, causing the valley to become narrow for the heavier fissioning nuclides, such as ^{252}Cf spontaneous fission [1–5]. The average mass numbers of the heavy products remain essentially unchanged near $A = 140$, while those of the light products increase linearly with increasing mass number of the fissioning nucleus [4].

The principal effect on mass-yield curves of increasing the excitation energy of a fissioning nucleus is to increase the probability of symmetric mass division, e.g., by two orders-of-magnitude for fission of ^{235}U by 14-MeV neutrons [1–3,5,6]. A decrease in excitation energy decreases

---

* Research supported by the National Science Foundation under Grant CHE-8003325.

NEW DIRECTIONS IN PHYSICS
The Los Alamos 40th Anniversary Volume
ISBN 0-12-492155-8

the probability of symmetric mass division; the decrease is by more than an order-of-magnitude for spontaneous fission of ^{240}Pu and ^{244}Cm; the bottoms of the valleys for these spontaneous-fission reactions have not been found [7,8].

For the heavier known spontaneous fissioning nuclides, e.g., isotopes of fermium, mass asymmetry still persists for the lighter isotopes, but for the heavier isotopes symmetric mass division predominates [9].

The large kinetic energy of primary fission fragments is thought to result mainly from Coulomb repulsion between fragments at scission. The kinetic energies increase as expected as the mass and atomic numbers of the light and heavy fragments become more nearly equal, until near mass number 130 an abrupt decrease in kinetic energy occurs [10,11]. This effect, amounting to $\sim$30 MeV and referred to as the "kinetic energy deficit", since Q values do not decrease in this mass region, is attributed to highly deformed fragments at scission. The decrease in kinetic energy occurs at about the same place that the decrease in mass yield occurs, near mass number 130.

The average total kinetic energy increases linearly with $Z^2/A^{\frac{1}{3}}$ of the fissioning nucleus [12], and this relationship continues for the heaviest nuclides, except for unusually high kinetic energies for ^{258}Fm and ^{259}Fm, which undergo spontaneous fission to give single symmetric mass-yield peaks [9]. However, the average total kinetic energy is normal for spontaneous fission of ^{259}Md, which also gives a single mass-yield peak [13].

Primary fission fragments are highly excited, due largely to their deformation at scission, and evaporate neutrons promptly after scission. Prompt neutron yields, symbol ν, vary with mass number to give the familiar "saw-tooth" distributions [14–16]. Neutron yields can be measured directly by neutron counting [15], or they can be deduced from mass-yield data by Terrell's summation method [14]. The results of the two methods are similar and show that near mass number 130 and $Z = 50$, the ν values are small, and, since kinetic energies are large, these fission fragments must be formed from compact scission configurations with little deformation for the heavy fragments.

Figure 13.1 shows yields of fission products from thermal-neutron-induced fission of ^{235}U. Mass yields, $Y(A)$, are shown on a linear scale at the back right. The independent or primary fission-product yields, labeled "IN", are plotted as a function of A and Z in the middle of the figure and resemble two jagged mountain ranges. This distribution of independent yields is the subject of this paper. It will be noted that the primary products are on the neutron excess side of beta stability, Z_A, so they undergo beta decay, which sums independent yields for each

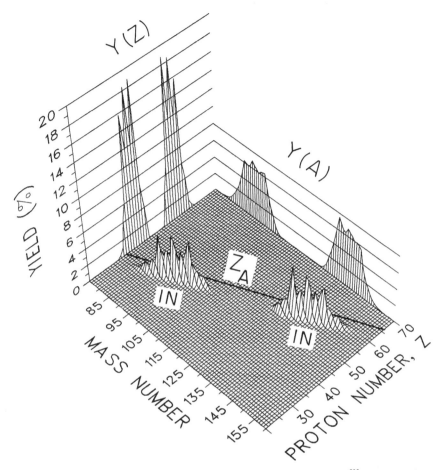

Figure 13.1. Yields of products from thermal-neutron-induced fission of ^{235}U. Independent yields, IN, are calculated from the A_P' model. The Z_A line indicates the approximate location of the most stable nuclides (beta stability).

mass number to give the cumulative yield of the last member of a decay chain, the mass yield, $Y(A)$.

The peaks in the independent-yield distribution for even Z's are believed to be associated with increased stabilization from proton pairing. The variation in yield for even and odd Z's is referred to as the even-odd Z effect, which also shows up in the isotopic yield curve, $Y(Z)$, obtained by summing independent yields over all A's for each Z, and shown at the back left of Figure 13.1. All significant independent yields must be measured or estimated for $Y(Z)$ sums to be calculated.

Early members of fission-product decay chains are very unstable and decay rapidly, with half-lives of a few seconds or less. A few of the beta transitions for early chain members populate highly excited daughter states, which emit neutrons. Beta-delayed neutron emission occurs for more than 100 fission products, and delayed-neutron yields depend on independent yields of early short-lived products far from beta stability, as well as on delayed-neutron emission probabilities, which are usually quite small. Total delayed-neutron yields amount to only ~1 per 100 fissions [17].

Experimental Methods

The more useful methods of measurement of independent yields are discussed briefly below. The radiochemical method accomplishes separation by Z, and appropriate mass numbers for the isotopes of a separated element can usually be deduced from decay properties, e.g., half-lives and gamma-ray energies. Chemical reactions used for separations can be very rapid; separation of phases is often the slow step, but use of automated equipment can minimize separation times [18,19]. Many separations can be accomplished in a few seconds. Therefore, independent yields of all but the shortest-lived fission products, those with half-lives of less than ~1 sec. can be measured, and, indeed, many yields have been determined by radiochemical methods.

The very rapid diffusion of Kr and Xe isotopes from powdered metal stearate salts allows separation times of ~0.1 sec., and yields of most Kr and Xe isotopes from a number of fission reactions have been measured radiochemically [20–22] and mass spectrometrically [23].

Mass spectrometers separate fission products by A; Z's are selected by differences in diffusion rates and in ionization-potentials of various elements. For example, Rb and Cs diffuse quite rapidly from hot graphite and are ionized on hot tantalum surfaces [24,25]. Relative yields for isotopes of Kr, Rb, Xe, and Cs have been accurately measured for a number of fission reactions; however, the relative yields require normalization to absolute yields measured in other ways, e.g., radiochemically.

A third method of measurement involves use of fission-product recoil separators: One named Lohengrin [26–28] is at Grenoble, and one named Hiawatha [29–31] is at Urbana. Separation times are ~1 μsec., so no beta decay occurs, an advantage over the previously described methods, which require corrections for beta decay. Yields for more than 100 light fission products from thermal-neutron-induced fission of ^{235}U and ^{233}U have been determined with recoil separators [26–28,30,31]. Measurements for other fission reactions are possible, and some measurements are in progress. Separation is by A/q, q being the ionic charge, and by kinetic energy, E_K; Z is determined by differential energy loss measurements, dE_K/dx, for which resolution is sufficient only for the lower-Z fission products, i.e., those that contribute to the light mass-yield peak. The independent yield of a fission product, $IN(A, Z)$, is distributed over many combinations of E_K and q, so summations must be made, requiring many measurements [26,30,31]. Also, approximations have been made with the assumption that yield values for average or most-probable kinetic energies and ionic-charge states are representative of the sums [27,28]. Small yield values, independent yields less than $\sim 0.1\,\%$, seem not to be reliable, since many are not consistent with data from other types of measurements, whereas large yield values generally are consistent [32,33].

Results and Discussion

Average values [33] of all measured independent yields of Rb and Cs isotopes from thermal-neutron-induced fission of ^{235}U are plotted on a logarithmic scale in Figure 13.2. Data for both elements are plotted against A', the sum of A and $\nu(A)$, the average number of neutrons emitted to form that A. Values of $\nu(A)$ were derived by use of a modification [32–35] of Terrell's mass-yield summation method [14]. The consistency of the two sets of experimental IN values shown on the plot indicates that the method of establishing mass-number complementarity is satisfactory for these and other elements formed in high yield.

Figure 13.3 shows average experimental independent yield data for complementary Mo and Sn isotopes and also recently measured yields for Tc and In isotopes [36,37], yields which are one to two orders-of-magnitude smaller than those of the Mo and Sn isotopes. It is noteworthy that the initial increase in the mass-yield curve, the dashed line, from the valley to the peaks is due almost entirely to the larger yields of Mo and Sn isotopes.

Yields of complementary elements ($Z_L + Z_H = 92$) from thermal-neutron-induced fission of ^{235}U are shown in Figure 13.4. The enhancement of yields of even Z elements over those of odd Z is evident. Estimates are required only for the very small element yields.

Yields for each neutron number are plotted in Figure 13.5. Data for the light peak are available for all higher yields, but estimates are required for many neutron numbers for the heavy peak, although the peak at $N = 82$ is based mainly on experimental data. The lack of a more

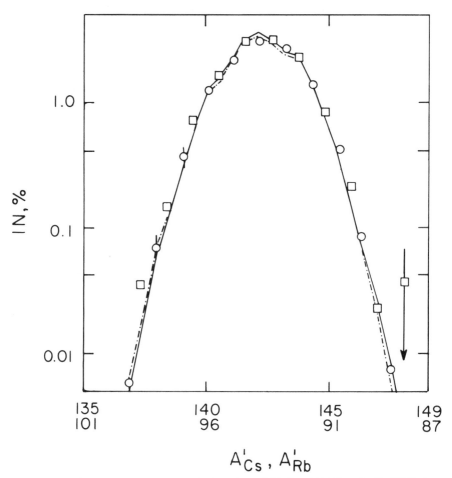

Figure 13.2. Average of experimental independent yields for $_{37}$Rb ($\square$) and $_{55}$Cs ($\bigcirc$) fission products from thermal-neutron-induced fission of ^{235}U. Lines are calculated from the Z_P model (--------) and the A_P' model (———).

pronounced even-odd neutron effect may be due, at least in part, to smoothing from prompt-neutron emission.

Figure 13.6 summarizes experimental data for thermal-neutron-induced fission of ^{235}U. Root-mean-square (RMS) and ΔZ values are plotted vs. A_H' or 236-A_L', ΔZ being the difference between the average Z and Z for unchanged-charge distribution (UCD). The lines are from the Z_P model, to be discussed, but without even-odd pairing effects. It can be seen that both the RMS and ΔZ values oscillate about the lines, an effect believed to be due mainly to the even-odd Z pairing effect. There is a

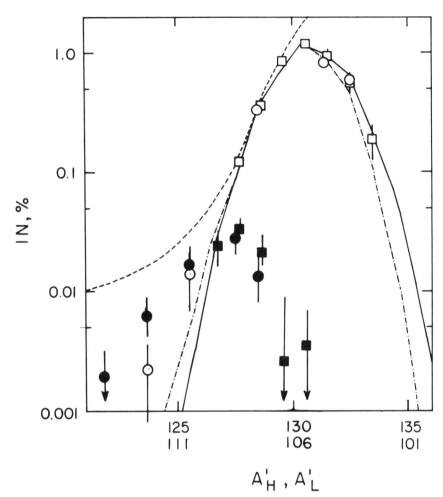

Figure 13.3. Average of experimental independent yields for $_{42}$Mo ($\square$), $_{50}$Sn ($\bigcirc$), $_{43}$Tc ($\blacksquare$), and $_{49}$In ($\bullet$) fission products from thermal-neutron-induced fission of ^{235}U. Lines are calculated from the Z_P model (·······) and the A'_P model (————) for $_{42}$Mo and $_{50}$Sn. The dashed line is a portion of the mass-yield curve, $Y(A)$ [3].

sharp increase in ΔZ near $Z = 50$ and also a dip in RMS values. The dashed lines near symmetric mass division (marked by X) are extrapolated; there are few data, so ΔZ and RMS values cannot be determined.

Calculations from the scission-point theory of Wilkins, Steinberg, and Chasman [38] are shown as filled triangles. The theoretical ΔZ values have the correct sign, show even-odd Z oscillations, and are close to

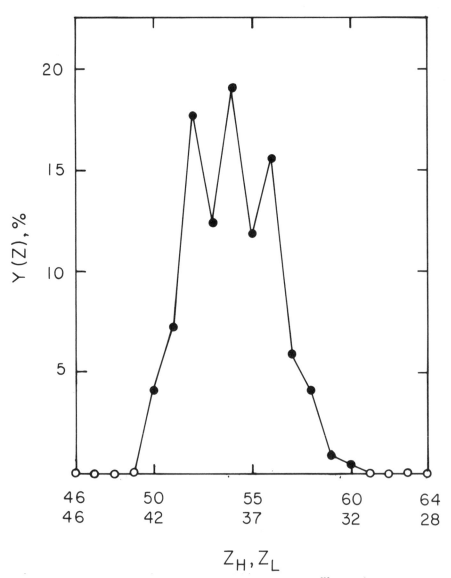

Figure 13.4. Element yields for thermal-neutron-induced fission of ^{235}U obtained by summing independent yields: ●, >90% of sum is from experimental independent yields; ○, >10% of sum is from Z_P model calculations.

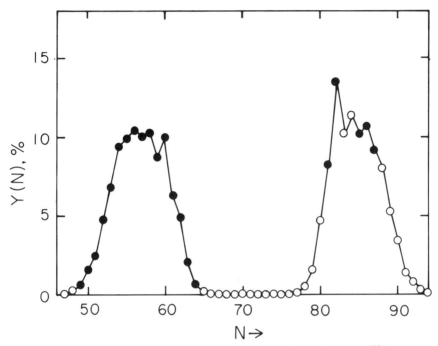

Figure 13.5. Yields by neutron number for thermal-neutron-induced fission of ^{235}U obtained by summing independent yields: ●, >90% of the sum is from experimental independent yields; ○, >10% of the sum is from Z_P model calculations.

experimental values just to the right of the $Z = 50$ line. To the far right the theoretical ΔZ values are too large and to the left, near symmetry, they are too small. The abrupt change in ΔZ near $Z = 50$, observed experimentally, is not reproduced by the theory. The average charge-dispersion width parameter, $\sigma_Z = 0.56$, determined in 1969 [35], was used to estimate a collective temperature of ~1.0 MeV, a value used in the theory, so there are no theoretical charge-dispersion widths for comparison with experimental values.

RMS and ΔZ values for thermal-neutron-induced fission of ^{233}U and ^{239}Pu and for spontaneous fission of ^{252}Cf show trends similar to those shown in Figure 13.6, but there are fewer data for the ^{239}Pu and ^{252}Cf fission reactions. Those that are available indicate that pairing effects are smaller than those observed for fission of ^{235}U and ^{233}U.

For all other fission reactions, there are even fewer nuclear-charge-distribution data than for the ^{239}Pu and ^{252}Cf fission reactions. In order to interpret the limited data available for most fission reactions, empirical

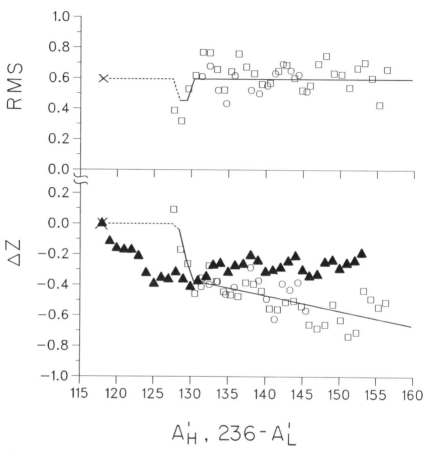

$$A'_H , 236 - A'_L$$

Figure 13.6. Experimental root-mean-square values, RMS, and the ΔZ function of average atomic numbers, $\bar{Z}$, for light ($\square$) and heavy ($\bigcirc$) products from thermal-neutron-induced fission of ^{235}U. $\Delta Z = Z_H - Z_{\text{UCD}} = Z_{\text{UCD}} - Z_L$, $Z_{\text{UCD}} = A'(92/236)$. Lines are from the Z_P model without even-odd pairing effects. The symbol $\blacktriangle$ represents calculated values, averaged over three A, from the scission-point theory [38]. Symmetric mass and charge division ($A' = 118$; $Z = 46$) is denoted by $\times$.

models can be formulated involving simple mathematical functions that are chosen to represent the data for thermal-neutron-induced fission of ^{235}U and ^{233}U. Then available data for other fission reactions can be used to determine values of parameters for the models as applied to each reaction.

All data for a fission reaction are treated together, i.e., the treatment is global. Gaussian dispersions of yields, modified by even-odd pairing effects, are assumed as functions of Z or of A'. Also, mass-number complementarity of light and heavy fission products are deduced from the average total number of neutrons, ν_T, evaporated to form a pair of fission products, a quantity deduced from mass yields by Terrell's summation method [14]. Values of ν_T are divided into ν_L and ν_H to give sawtooth functions similar to those observed experimentally [14–16]. The mass-number complementarity relationship, equation (1), is only approximate, since $\nu(A)$ values are averages.

$$A'_L + A'_H \simeq A_{\text{FIS}}, \qquad A' = A + \nu(A) \tag{1}$$

However, atomic-number complementarity is essentially exact, since very few light charged particles are emitted in low-energy fission, so complementary element yields should be equal.

Once parameters are determined, e.g., by the method of least squares, independent yields can be calculated for all fission products, as well as their sums over atomic, neutron, and mass numbers.

Various modifications of the Z_P model have been widely used [20–22,32–35,39–42]. Fractional independent yields

$$\text{FI}(Z, A) = \frac{\text{IN}(Z,A)}{Y(A)} \tag{2}$$

are currently calculated on the assumption that their dispersion with Z at constant A is Gaussian. FI(Z,A) is the product of a Gaussian function of $Z_P(A)$ and $\sigma_Z(A)$, an even-odd factor, EOF(A), and a normalization factor, NF(A). $Z_P(A)$, the "most probable charge", is the Z value, usually fractional, at which the Gaussian function is maximum. $Z_P(A)$ is determined from a ΔZ function and Z for unchanged charge distribution (UCD).

$$Z_P(A) = Z(A)_{\text{UCD}} + \Delta A(A), \qquad Z(A)_{\text{UCD}} = A'(Z_{\text{FIS}}/A_{\text{FIS}}) \tag{3}$$

The width parameter, $\sigma_Z(A)$, is slightly less than RMS because of Sheppard's grouping correction.

$$\sigma_Z(A) = \{[\text{RMS}(A)]^2 - \tfrac{1}{12}\}^{\frac{1}{2}} \tag{4}$$

EOF(A) is the product of EOZ(A) and EON(A), the even-odd proton and neutron factors, for even-even nuclides, and it is the reciprocal of this product for odd-odd nuclides. EOF(A) is the ratio of EOZ(A) to EON(A) for even-odd nuclides, and it is the reciprocal of this ratio for odd-even nuclides. The normalization factor, NF(A), insures that the sum of fractional independent yields is unity for each A. (Application of the even-odd factors disturbs the inherent normalization of a Gaussian area by several percent.) The data used to derive Z_P model parameters are fractional independent and fractional cumulative yields, provided the latter are not close to unity. Equations used for the Z_P model calculations are given in reference 34.

The Z_P model has been very useful for interpretation of independent yield data for nuclides contributing to the mass-yield peaks, and these yields constitute most available experimental data. However, as will be discussed, problems arise in representing the data available for nuclides in the region where mass yields change rapidly in going from the valley to the peaks (A_H = 120–130). Therefore, another model has been proposed [32], the A_P' model, which may avoid these problems and gives a different perspective of nuclear-charge distribution in fission. It will be compared with the Z_P model.

In the A_P' model independent yields, $IN(A,Z)$, are calculated on the assumption that their dispersion with A' at constant Z is Gaussian. $IN(A,Z)$ is the product of an element yield, $Y(Z)$, a Gaussian function of $A_P'(Z)$ and $\sigma_{A'}(Z)$, an even-odd factor, $EOF(Z)$, and a normalization factor, $NF(Z)$. $A_P'(Z)$, "the most probable average mass number of the precursor fragments", is the A' value at which the Gaussian function is maximum; $\sigma_{A'}(Z)$ is a width parameter; and $EOF(Z)$ refers only to the even-odd neutron effect, since even-odd proton effects are included in the $Y(Z)$ values. Data for complementary elements ($Z_L + Z_H = Z_{FIS}$) are treated together to give equal $Y(Z_L)$ and $Y(Z_H)$ values. $A_P'(Z)$ is determined from a $\Delta A'$ function and A' for unchanged charge distribution.

$$A_P'(Z) = A'(Z)_{UCD} + \Delta A'(Z), \qquad A'(Z)_{UCD} = Z\left(\frac{A_{FIS}}{Z_{FIS}}\right) \qquad (5)$$

Data used to derive A_P' model parameters include both independent yields and mass yields, which are sums of independent yields. Equations used for the A_P' model calculations are given in the Appendix.

The A_P' model has the feature of treating independent and mass yields together by a least-squares procedure. Thus, once the charge-distribution parameters ($Y(Z)$, $A_P'(Z)$, $\sigma_{A'}(Z)$, $EON(Z)$) are known or deduced from systematic trends, a partial set of $Y(A)$ values can be used to calculate a complete mass-yield curve, as well as a complete set of independent yields.

The $\nu(A)$ vs. A functions used for calculating A' ($A' = A + \nu(A)$) are derived from ν_T, the average total number of neutrons emitted to form complementary product pairs, and calculated by Terrell's method [14] from Rider's 1981 mass-yield data [3]. The methods of dividing ν_T between ν_L and ν_H and of extrapolating these values into mass-number regions for the valley and for the wings of a mass-yield curve are discussed in reference 34.

The parameters derived from least-squares calculations with the Z_P model are shown for four fission reactions in Table 13.1. Average $\overline{\sigma_Z}$ values were determined, and there is an indication that these values increase with the A and Z of the fissioning nucleus. The ΔZ function, assumed to be linear over the A regions for mass-yield peaks, is essentially constant at $A' = 140$ (~ -0.5), and it has a small negative slope. Average even-odd factors are much larger for protons than for neutrons, and both decrease considerably with increasing A and Z of the fissioning nucleus and also with excitation energies of only several MeV, an effect not shown in the table, but for which considerable data exist [34,43].

Figures 13.7 and 13.8 show $\Delta A'$ functions and $Y(Z)$ values, respectively, calculated for the A_P' model by the method of least-squares. A table of the other parameters derived is given in the Appendix. The $\Delta A'$ values for $Z > 50$ can be represented as a function of $Y(Z)$, but those for $Z < 49$ had to be estimated because of the lack of data. $\Delta A'$ is a maximum at $Z = 50$ for all four fission reactions. The average width parameter, $\overline{\sigma_{A'}}$, may increase a little with the A and Z of the fissioning nucleus. The even-odd neutron factors are close to unity, as found with the Z_P model.

The results from Z_P- and A_P'-model calculations are compared in Figures 13.2 and 13.3 with experimental data for thermal-neutron-induced fission of ^{235}U. As shown in Figure 13.2, both model calculations represent the Rb and Cs data well, a typical result for high-yield elements for which many data exist.

Significant differences appear between the two model calculations for Mo and Sn yields, shown in Figure 13.3; calculated values from the A_P' model agree better with several of the higher yields than do those from the Z_P model.

Independent yields calculated from the two models are compared with measured Tc and In yields [36,37] in Figure 13.9. It can be seen that the A_P' model represents the data well. The curve from the Z_P model calculations has a strange shape and is inconsistent with a number of

Table 13.1. Z_p model parameters.

	$\overline{\sigma_Z}$	$\Delta Z(A' = 140)$	$\dfrac{d\Delta Z}{dA_H'}$	$\overline{EOZ}$	$\overline{EON}$
^{235}U + n_{th}	0.52	-0.47	-0.010	1.27	1.08
^{233}U + n_{th}	0.54	-0.50	-0.015	1.26	1.05
^{239}Pu + n_{th}	0.55	-0.48	-0.015	1.06	1.00
^{252}Cf S.F.	0.61	-0.47	-0.005	1.00	1.00
(Estimated uncertainty)	(± 0.02)	(± 0.02)	(± 0.005)	(± 0.03)	(± 0.03)

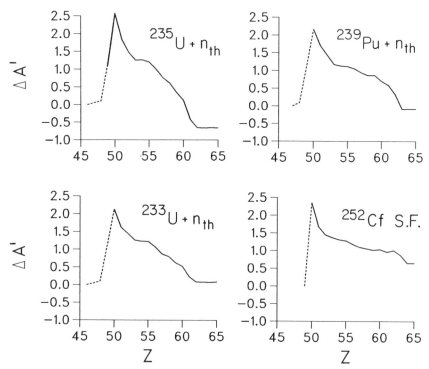

Figure 13.7. $\Delta A'$ functions for the A'_P model. A portion of the functions were derived from experimental data (———), and a portion was assumed (-------).

the data; there is no apparent way to correct these defects. It appears that the basic assumption of a Gaussian dispersion in Z for each A is not valid for this mass-number region, where mass yields change rapidly.

Figure 13.10 shows comparisons of experimental RMS and average Z values, expressed as ΔZ, with values calculated from the two models for thermal-neutron-induced fission of ^{235}U. It can be seen that the two models represent the data reasonably well and about equally well. Oscillations in the functions due to even-odd pairing effects are evident, and the experimentally observed decrease in RMS and the sharp rise in ΔZ near $Z = 50$ are reproduced by both empirical models. The models give somewhat different predictions near symmetric mass division; however, no data exist from which model parameters could be derived or to which model predictions could be compared.

Another test of models for nuclear-charge distribution in fission is detailed charge balance or conservation, i.e., the equality of yields of

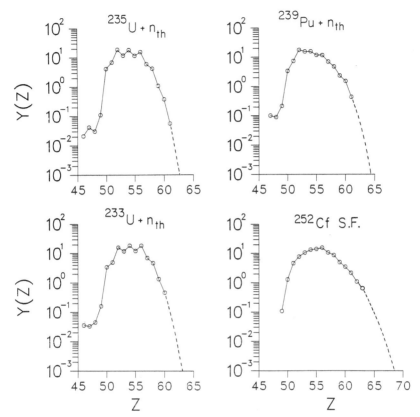

Figure 13.8. $Y(Z)$ values determined for the A'_P model. Many individual values were determined (—⊖—⊖—); those on the wing (------) were determined from a Gaussian function described in the Appendix.

complementary elements. Several models are compared in Figure 13.11, which shows the difference from unity of ratios of yields for complementary light and heavy elements as a function of Z_H for thermal-neutron-induced fission of ^{235}U. The charge balance for the A'_P model is perfect (the line), since equality of yields of complementary light and heavy products is an inherent feature of the model. The charge balance is also quite good for most peak yields, $Z_H = 50–60$, for the Z_P model. Element yield ratios differ by as much as 10% near symmetric charge division.

Other treatments of nuclear-charge distribution in fission, e.g., Evaluated Nuclear Data File B-V [44] and Crouch's compilation [2], give deviations of yield ratios from unity that are often quite large.

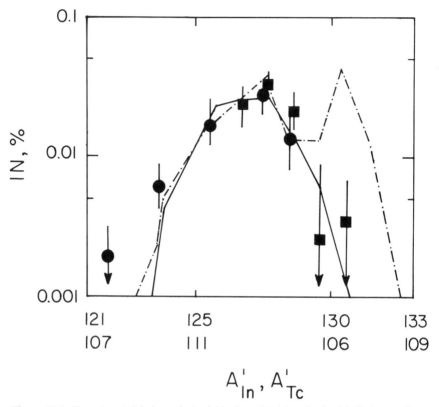

Figure 13.9. Experimental independent yields for $_{43}$Tc (■) and $_{49}$In (●) fission products from thermal-neutron-induced fission of ^{235}U. Lines are calculated from the Z_P model (·······) and the A'_P model (———).

Another test of the models is comparison of measured delayed-neutron yields with those calculated from a model. As shown in Table 13.2, the agreement between total yields is satisfactory in that calculated values are equal to or somewhat less than measured values, as would be expected from use in the calculations of only 48 of the higher delayed-neutron yields, those for fission products with known delayed-neutron emission probabilities, P_n, of more than 100 known delayed-neutron precursors. The agreement is evidence that the estimates of independent yields far from beta stability are reasonable.

Figure 13.12 shows a comparison of mass-yield curves from Rider's compilation [3] (the points) and from A'_P model calculated values (the

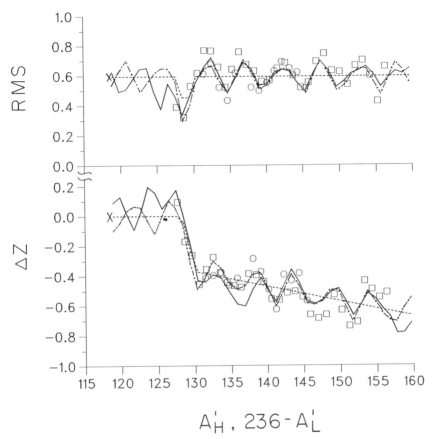

Figure 13.10. RMS and ΔZ functions calculated from the Z_P model (------) and A_P' model (———) compared to the same experimental data ($\square$, $\bigcirc$) shown in Figure 13.6.

heavy line) for thermal-neutron-induced fission of ^{233}U. The calculated curve results from summation for each mass number of the independent yields plotted for each element (light solid and dashed lines). The atomic numbers are given near the bottom of the figures just above the mass numbers; yields are plotted on a logarithmic scale. The calculated independent yield curves for tin ($Z = 50$) and its complement, molybdenum ($Z = 42$) are marked with X's. It can be seen, as discussed for thermal-neutron-induced fission of ^{235}U (Fig. 13.3), that most of the initial rise in yield in going from the valley to the peaks is due to larger yields of $_{50}$Sn and $_{42}$Mo isotopes compared to yields of $_{49}$In and $_{43}$Tc.

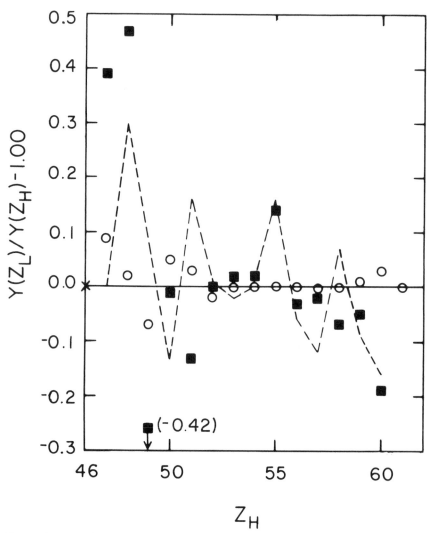

Figure 13.11. Deviations of various models from elemental charge balance.————, A'_P model; ○, Z_P model; ■, ENDF/B-V recommended values [44];— — —, Crouch's recommended values [2].

The calculated mass-yield curve represents the experimental mass-yield data well, except near the valley. The experimental data were used in the calculation, so the agreement achieved indicates only that the empirical A'_P model offers a good correlation of mass-yield and nuclear-charge-distribution data. It also provides a means of estimating unmeasured yields, both independent and mass yields.

Table 13.2. Total delayed-neutron yields. (No. per 100 fissions)

	Measured (All, >102)	Calculated (48 known P_n)	
		Z_P Model	A_P' Model
^{235}U + n_{th}	1.62 ± 0.05	1.44	1.55
^{233}U + n_{th}	0.67 ± 0.03	0.69	0.72
^{239}Pu + n_{th}	0.63 ± 0.04	0.61	0.66
^{252}Cf S.F.	0.89 ± 0.30	0.70	0.61
		(Estimated uncertainty ≈ 10%)	

Figure 13.13 shows the same type of plot for thermal-neutron-induced fission of ^{239}Pu. Again the fit of the experimental points by the calculated curve is good, except near the valley. The contribution of independent yields of $_{50}$Sn and $_{44}$Ru isotopes to the initial rise in mass yields from the valley to the peaks is also evident in this plot. Thus it is $Z = 50$,

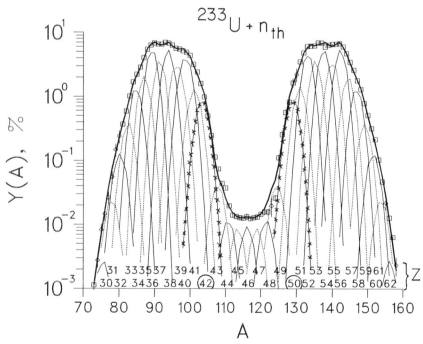

Figure 13.12. Mass- and independent-yield curves for products from thermal-neutron-induced fission of ^{233}U. Mass yields: □, experimental [3]; ◇, estimated [3]; ▬, A_P' model values. Calculated elemental independent-yield curves:—, even Z; ·········, odd Z; ✗✗✗, $Z = 42$ and $Z = 50$.

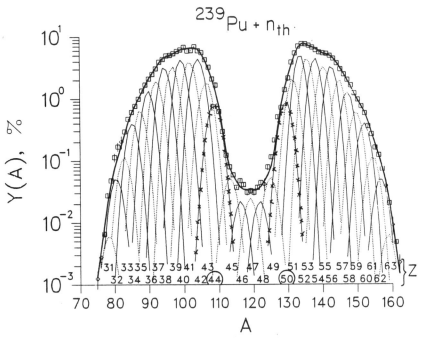

Figure 13.13. Mass- and independent-yield curves for products from thermal-neutron-induced fission of ^{239}Pu. Symbols are the same as for Figure 13.12, except ✖✖✖ is for $Z = 44$ and $Z = 50$.

not the charge of the complement, that is responsible for the initial rise in mass yields.

Future Investigations

Future investigations of nuclear-charge distribution in fission should include additional experimental measurements, extension and refinement of theoretical treatments, and further correlations of nuclear-charge distribution systematics with other properties of nuclear-fission reactions.

Experimental techniques should be refined and new ones developed to allow measurements of independent or fractional-independent yields for short-lived, low-yield fission products from near symmetric fission and from very asymmetric fission. Recent independent yield measurements for $_{43}$Tc and $_{49}$In fission products [36,37] (Fig. 13.9) are interesting and important contributions in this direction. Also, current techniques, as well as refined or new ones, should be applied to the many fission reactions for which no, or only a few, independent-yield data exist. Additional data correlated and evaluated by use of empirical models, such as those that have been discussed, could establish additional systematic trends in fission-product yields with atomic and mass numbers and with excitation energy of fission products and of fissioning nuclei. Knowledge of additional systematic trends would improve the reliability of model predictions of the great many unmeasured fission yields. Also, systematic trends could be used to test theoretical predictions based on various physical models and could serve as a basis for choosing between the physical models, thus increasing our understanding of the mechanism of nuclear fission.

The scission-point theory [38] attributes asymmetric charge and mass division mainly to the influence of deformed neutron shells. The calculated charge and mass distributions are qualitatively consistent with many experimental observations, although quantitative yield predictions are often poor. For example, as has been discussed, the theory does not account for the observed abrupt yield change near the 50-proton shell. Theoretical treatments based on other physical models, e.g., saddle-point, liquid-drop, and statistical theories, have been less successful than the scission-point theory in accounting for charge and mass distributions, often because complexities of the calculations do not allow detailed fission-yield predictions to be made. It is hoped that rapid, large-memory computers will allow refinements and extensions of the various theories so that detailed predictions can be made for comparison with experiments.

As an illustration of what might be learned from consideration of several fission properties together, it is interesting to speculate about possible nuclear-charge distributions for spontaneous fission of ^{258}Fm and ^{259}Md, two of the heaviest nuclides for which kinetic-energy and mass-yield distributions have been measured, but for which no nuclear-charge-distribution data are available. Symmetric mass division has been reported for both reactions; fragments from ^{259}Md have a normal average total kinetic energy [13] whereas fragments from ^{258}Fm have an average total kinetic energy $\sim$40 MeV higher than normal [9].

The A'_P model can be used to calculate element yields and their mass-number dispersion from mass-yield data, provided that charge-distribution parameters can be deduced from those determined for lighter nuclides, e.g., ^{252}Cf spontaneous fission. Calculations for ^{258}Fm and ^{259}Md were made assuming $\overline{\sigma_{A'}} = 1.60$ and $\overline{EON} = 1.00$, the values determined for ^{252}Cf. The assumed $\Delta A'$ function for $Z_H > 50$ was a linear function, $\Delta A' = 1.25 - 0.05(Z_H - 50)$, approximating the function for ^{252}Cf shown in Figure 13.7. From symmetry considerations, $\Delta A'(50) = 0.0$ for ^{258}Fm, and $\Delta A'(50) = -1.20$ for ^{259}Md. (Calculations were also made with

$\Delta A' = 0.0$ for all elements, and results similar to those discussed below were obtained.)

The calculated yields for ^{258}Fm are plotted on a linear scale in Figure 13.14 along with experimental fragment-mass yields (points). It can be seen that our treatment indicates that the sharp mass-yield peak is due almost entirely to the formation of $_{50}$Sn isotopes; about one-half of the fissions give two tin nuclei ($Y(Z = 50) \simeq 100\%$ of the 200% total). For these nearly spherical nuclei, compact scission configurations are expected and would lead to very high kinetic energies. Even after averaging these with lower kinetic energies associated with isotopes of other elements, formed in low yield, the average total kinetic energy would be large, as observed [9].

The calculated and experimental yields for ^{259}Md, also plotted on a linear scale, are shown in Figure 13.15. The calculations indicate that mass asymmetry is possible; also of interest is the broad distribution of element yields. The latter results largely from the broad experimental mass-yield curve, and suggests that many scission configurations, some compact and some deformed, contribute to the distributions in fragment

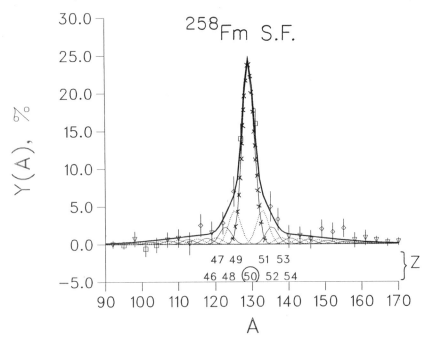

Figure 13.14. Mass- and independent-yield curves for fragments from spontaneous fission of ^{258}Fm. Symbols are the same as for Figure 13.12, except ✕✕✕ is for $Z = 50$.

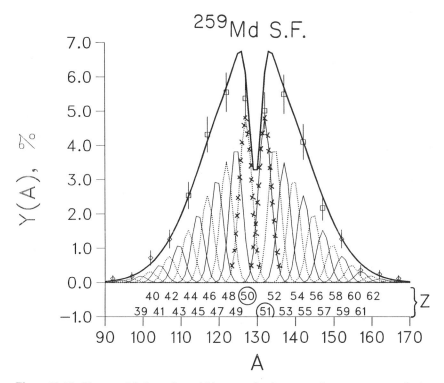

Figure 13.15. Mass- and independent-yield curves for fragments from spontaneous fission of ^{259}Md. Symbols are the same as for Figure 13.12, except ⋯⋯ is for even Z, ——— is for odd Z, and ·×·×·×· and ✕✕✕ are for $Z = 50$ and 51, respectively.

mass and atomic numbers and in kinetic energy (A_f, Z, and E_K). A normal average total kinetic energy, such as obsrved, would result from averaging over a broad E_K distribution, which is typical for most fission reactions. In fact, a much broader mass distribution was observed for fragments with low kinetic energies than for those with high kinetic energies from spontaneous fission of ^{259}Md [13].

It is interesting that the difference of a single proton results in a large difference in the mass-yield distributions determined experimentally for the ^{258}Fm and ^{259}Md spontaneous fission reactions, and probably also, as calculated, in the charge and independent yield distributions. It is also interesting that for fission of many lighter nuclei an abrupt increase in product yields occurs when Z is increased from 49 to 50, an increase that may be associated with the nearly constant position of the heavy mass-yield peak. These observations, along with the sensitivity of the even-odd proton effect on Z, A, and excitation energy of a fissioning nucleus, indicate that single-particle effects are important in determining the course of fission reactions. Understanding of how and where from saddle to scission this influence occurs will probably require both experimental and theoretical research in "new directions" as well as extensions and refinement of more conventional methods.

Acknowledgments

It is a pleasure for the author to acknowledge valuable contributions to the development of computer programs by H. Dworsky, E. Plachy, D. Roman, and K. Wong. He also gratefully acknowledges valuable assistance in the preparation of illustrations by H. Dworsky and R. Holdzkom.

Appendix

The equations for the A_P' model given below, involving the error function of x, ERF(x), are incorporated in a computer program for use with a modification of the general least-squares program, ORGLS [45].

$$IN(A,Z) = [0.5][Y(Z_H)][EOF][NF(Z)][ERF(VA) - ERF(WA)]$$

$$VA = \frac{A' - A_P'(Z) + 0.5}{\sigma_{A'}(Z)\sqrt{2}}$$

$$WA = \frac{A' - A_P'(Z) - 0.5}{\sigma_{A'}(Z)\sqrt{2}} \tag{6}$$

$$Y(Z_L) = Y(Z_{H_c}), \qquad Z_L + Z_{H_c} = Z_{FIS}$$

Individual $Y(Z_H)$ values are determined, or, for the larger Z_H values, the following Gaussian function is used:

$$Y(Z_H) = [0.5][R][EOZ][ERF(VY) - ERF(WY)]$$

$$VY = \frac{ZM - Z_H + 0.5}{\sigma_Y\sqrt{2}}$$

$$WY = \frac{ZM - Z_H - 0.5}{\sigma_Y\sqrt{2}} \tag{7}$$

$EOZ = \overline{EOZ}$ for even Z

$EOZ = 1/\overline{EOZ}$ for odd Z

$EOF = \overline{EON}$ for even N

$EOF = 1/\overline{EON}$ for odd N

$A' = A + \nu(A)$

$A'_P(Z_L) = Z_L(A_{FIS}/Z_{FIS}) - \Delta_{A'}(Z_{H_c})$

$A'_P(Z_H) = Z_H(A_{FIS}/Z_{FIS}) + \Delta_{A'}(Z_H)$

$\Delta_{A'}(Z_H > 50) = C + 0.5[S]\{\ln[Y(Z_H + 1)] - \ln[Y(Z_H - 1)]\}$

$\Delta_{A'}(Z_H = 48,49,50)$ Individual values are determined, or fixed values are assumed. $\quad$ (8)

$$\Delta_{A'}(Z_H < 48) = \frac{\Delta_{A'}(48)}{\left(48 - \dfrac{Z_{FIS}}{2}\right)}$$

$\overline{\sigma}_{A'}$ is determined, and some of the individual values of $\sigma_{A'}(Z)$ may be determined and/or fixed values assumed.

A normalization factor, NF(Z), is required for each Z because the even-odd factor destroys the inherent normalization properties of Gaussian distributions and because the areas of unit A' width, which represent independent yields, are not contiguous because of variation of $\nu(A)$ with A. The NF(Z) values seldom deviate from unity by more than 10%.

Parameters for the A'_P model are shown in Table 13.3; values with errors were determined by least-squares calculations; values without errors were assumed and not allowed to vary during the calculations. The $Y(Z)$ values determined are shown in Figure 13.8, and the $\Delta A'$ functions are shown in Figure 13.7.

Table 13.3. Parameters for the A'_P model.

Parameter	^{235}U + n$_{th}$	^{233}U + n$_{th}$	^{239}Pu + n$_{th}$	^{252}Cf S.F.
$\overline{\sigma}_{A'}$	1.47 ± 0.02	1.52 ± 0.02	1.53 ± 0.03	1.60 ± 0.04
C	1.27 ± 0.03	1.22 ± 0.03	1.20 ± 0.05	1.26 ± 0.07
S	0.79 ± 0.04	0.57 ± 0.05	0.52 ± 0.09	0.92 ± 0.18
$\Delta_{A'}(50)$	2.57 ± 0.06	2.13 ± 0.08	2.16 ± 0.13	2.40 ± 0.23
$\Delta_{A'}(49)$	1.1 ± 0.2	1.1	1.1	0.0
$\Delta_{A'}(48)$	0.1	0.1	0.1	—
EON	1.06 ± 0.02	1.06 ± 0.02	1.02 ± 0.03	1.00
Following are $Y(Z)$ parm.				
for $Z \geqslant$	62	62	63	64
R	100.	100.	100.	50.[a]
ZM	54.2[a]	54.3[a]	54.2[a]	57.0[a]
σ_Y	1.87 ± 0.01	1.96 ± 0.01	2.30 ± 0.02	2.69 ± 0.03
EOZ	1.27	1.26	1.08	1.00

[a] Determined in previous calculations in which the Gaussian function was used to determine $Y(Z_H)$ for many Z_H, including $Y(Z)$ values on and/or near the peak.

References

[1] Katcoff, S., (1960) *Nucleonics* **18**(11), 201.

[2] Crouch, E.A.C., (1977) *Atomic Data and Nuclear Data Tables* **19**(5), 417.

[3] Rider, B.F., (1981) "Compilation of Fission Product Yields", *Vallecitos Nuclear Center Report* No. NEDO-12154-3(c).

[4] Flynn, K.F., Horwitz, E.P., Bloomquist, C.A., Barnes, R.F., Sjoblom, R.K., Fields, P.R., and Glendenin, L.E., (1972) *Phys. Rev. C* **5**, 1725.

[5] Wahl, A.C., (1965) in *Physics and Chemistry of Fission*, Vol. I, IAEA, Vienna, 317.

[6] Glendenin, L.E., Gindler, J.E., Henderson, D.J., and Meadows, J.W., (1981) *Phys. Rev. C* **24**, 2600.

[7] Cowan, G.A., Bayhurst, B.P., Prestwood, R.J., Gilmore, J.G., and Sattizahn, J.E., (1973) *Personal communication*, unpublished.

[8] Flynn, K.F., Srinivasan, B., Manuel, O.K., and Glendenin, L.E., (1972) *Phys. Rev. C* **6**, 2211.

[9] Hoffman, D.C., (1980) in *Physics and Chemistry of Fission*, Vol. II, IAEA, Vienna, 275.

[10] Neiler, J.N., Walter, F.J., and Schmitt, H.W., (1966) *Phys. Rev.* **149**, 894.

[11] Brissot, R., Bocquet, J.P., Ristori, C., Crancon, J., Guet, C.R., Nifenecker, H.A., and Montoya, M., (1980) in *Physics and Chemistry of Fission*, Vol. II, IAEA, Vienna, 99.

[12] Viola, V.E., Jr., (1966) *Nucl. Data Sect. A* **1**, 391.

[13] Wild, J.F., Hulet, E.K., Lougheed, R.W., Baisden, P.A., Landrum, J.H., Dougan, R.J., and Mustafa, M.G., (1982) *Phys. Rev. C* **26**, 1531.

[14] Terrell, J., (1962) *Phys. Rev.* **127**, 880.

[15] Nifenecker, H., Signarbieux, C., Babinet, R., and Poitou, J., (1974) in *Physics and Chemistry of Fission*, Vol. II, IAEA, Vienna, 117.

[16] Gindler, J., (1979) *Phys. Rev. C* **19**, 1806.

[17] Rudstam, R., (1978) "Status of Delayed Neutron Data", *Proceedings of the Petten 1977 Conference on Fission Product Nuclear Data*, IAEA-213, Vol. II, 567.

[18] Herrmann, G., and Trautmann, N., (1982) *Ann. Rev. Nucl. and Part. Sci.* **32**, 117.

[19] Meyer, R.A., (1979) *Lawrence Livermore Lab. Preprint* UCRL-81069.

[20] Wahl, A.C., (1958) *J. Inorg. Nucl. Chem.* **6**, 263.

[21] Wahl, A.C., Ferguson, R.L., Nethaway, D.R., Troutner, D.E., and Wolfsberg, K., (1962) *Phys. Rev.* **126**, 1112.

[22] Wolfsberg, K., (1965) *Phys. Rev.* **137**, B929.

[23] Brissot, R., Crancon, J., Ristori, Ch., Bocquet, J.P., and Moussa, A., (1975) *Nucl. Phys.* **A255**, 461.

[24] Balestrini, S.J., and Forman, L., (1975) *Phys. Rev. C* **12**, 413.

[25] Balestrini, S.J., Decker, R., Wollnik, H., Wünsch, K.D., Jung, G., Koglin, E., and Siegert, G., (1979) *Phys. Rev. C* **20**, 2244.

[26] Lang, W., Clerc, H.-G., Wohlfarth, H., Schrader, H., and Schmidt, K.-H., (1980) *Nucl. Phys.* **A345**, 34.

[27] Clerc, H.-G., Lang, W., Wolfarth, H., Schmidt, K.-H., Schrader, H., Pferdekämper, K.E., and Jungmann, R., (1975) *Z. Physik A* **274**, 203.

[28] Wollnick, H., Siegert, G., Grief, J., and Fiedler, G., (1976) *CERN Report on Exotic Nuclei*, Corsica, France; (1976) *Phys. Rev. C* **14**, 1864.

[29] Dilorio, G., and Wehring, B.H., (1976) *Trans. Am. Nucl. Soc.* **23**, 523.

[30] Strittmatter, R.B., and Wehring, B.H., (1978) *Personal communication*; Strittmatter, R.B., (1978) *Ph.D. thesis*, Univ. of Illinois, Urbana.

[31] Wehring, B.H., Lee, S., and Swift, G., (1980) *Univ. of Illinois Nucl. Rad. Lab. Report* No. UILU-ENG-80-5312, unpublished.

[32] Wahl, A.C., (1978) *IAEA Report* No. INDC(NDS)-87, Vienna, 215.

[33] Wahl, A.C., (1982) *Compilation and evaluation of nuclear-charge-distribution data for thermal-neutron-induced fission of* ^{235}U, ^{233}U, *and* ^{239}Pu *and for spontaneous fission of* ^{252}Cf, unpublished.

[34] Wahl, A.C., (1980) *J. Radioanal. Chem.* **55**, 111.

[35] Wahl, A.C., Norris, A.E., Rouse, R.A., and Williams, J.C., (1969) in *Physics and Chemistry of Fission*, IAEA, Vienna, 813.

[36] Trautmann, N., and Fassbender, T., (1980) *Personal communication*; Fassbender, T., (1979) *Diplomarbeit*, Univ. of Mainz.

[37] Semkow, T.M., (1983) *Ph.D. thesis*, Washington Univ., St. Louis.

[38] Wilkins, B.D., Steinberg, E.P., and Chasman, R.R., (1976) *Phys. Rev. C* **14**, 1832.

[39] Glendenin, L.E., Coryell, C.D., and Edwards, R.R., (1951) Paper 52 in *Radiochemical Studies: The Fission Products*, Coryell, C.D., and Sugarman, N., Editors, McGraw-Hill, Book 1, 489.

[40] Coryell, C.D., Kaplan, M., and Fink, R.D., (1961) *Can. J. Chem.* **39**, 646.

[41] Nethaway, D.R., (1974) *Lawrence Livermore Laboratory Report* No. UCRL-51538.

[42] Wolfsberg, K., (1974) *Los Alamos Lab. Report* No. LA-5553-MS.

[43] Mariolopoulos, G., Hamelin, Ch., Blachot, J., Bocquet, J.P., Brissot, R., Crancon, J., Nifenecker, H., and Ristori, Ch., (1981) *Nucl. Phys.* **A361**, 213.

[44] England, T.R., (1981) *Personal communication* of independent fission yields from ENDF/B-V.

[45] Busing, W.R., and Levy, H.A., (1962) *Oak Ridge National Laboratory Report* No. ORNL-TM-271.

14 Developing Larger Software Systems

Eldred Nelson

Introduction

During the 40 years since the founding of Los Alamos Laboratory in April, 1943, information processing automation has expanded from limited use of hand calculators and punched card machines to extensive use of digital computers throughout government and industry. This expansion was driven by the development of the electronic digital computer, continually decreasing cost of digital hardware, and continually improving tools and techniques supporting the application of computers to increasingly complex information processing tasks. As the next 40 years begin, new digital architectures supporting parallel and distributed computing, data communication linking computers on both local and global bases, and personal computers usable in the home and office are providing vast new information processing capabilities potentially applicable to a new dimension of complex problems important to national defense, economic growth, and international competitiveness. Expansion of information processing automation into the domain of these problems is requiring new directions in information processing software technology. Progress to meet the challenge of system complexity surpassing that in current large systems is being made in two directions: integrated software tool sets and a formal model of software structure supporting the management of complex system

NEW DIRECTIONS IN PHYSICS
The Los Alamos 40th Anniversary Volume
ISBN 0-12-492155-8

development. Other techniques expected to contribute to developing larger software systems, such as artificial intelligence, are not discussed in this paper.

The Challenge of Large Systems

Because the development of current large systems has strained available resources and techniques, some have questioned the capability to successfully develop larger more complex systems. Principal problems in managing the development of large systems are:

* Developing and maintaining a coherent picture of the system and its development status;
* Defining and keeping track of the large number of relations among the system elements, system documentation, and system development personnel;
* Keeping the relations up-to-date as changes in individual elements require changes in other elements.

These problems become increasingly severe as systems become larger and larger, and the effects of these problems are well known: difficulty in meeting performance objectives, schedule slippage, and cost overruns. Can these problems be kept in bounds as the system size increases? Is there a need for more complex systems? Answers to these questions have been sought by looking at the history of information processing system development and the nature of information system complexity.

At each stage in the evolution of electronic digital computers, applications straining the capacity of the digital hardware and software development techniques have arisen, creating demands for more capability. Even the earliest computers were used in demanding applications. The first problem run on the first electronic digital computer, the ENIAC, was a hydrogen bomb design problem [1]. The first commercial digital computer, the UNIVAC I, began processing US Census data in April, 1950. In the mid-1950s, digital computers began to be used extensively in defense system development, supporting aircraft design and ballistic missile development. Nuclear reactor design and weather forecasting were other large applications initiated during this period. With the introduction of solid state computers in 1960 and third generation (multiprogramming) computers in 1965, the increased information processing power led to increasingly complex applications—e.g., supporting manned space flights and the moon landings.

In the early 1970s, when development of ballistic missile defense systems entered its second phase, some questioned whether the complexity of the software, required in such systems to adequately defend against a barrage of ballistic missiles, was beyond the capability of people to design, check-out, test, and assure its correct operation. These fears turned out to be unjustified, for new software development techniques and tools (structured programming, programming standards, code auditing, and automated test tools) combined with an effective new software development management discipline were successfully applied [2,3] to ballistic missile defense in the Site Defense Program.

Now the same question has been raised following the President's announcement of increased emphasis on ballistic missile defense, for today's threat may involve a barrage of a thousand missiles, each accompanied by decoys and/or other penetration aids. Identification and tracking of the missiles and direction of their interception will require very elaborate data processing. Other demanding applications, some of them extensions of old problems and others required to solve new problems, may be found in business and government. Increasing the accuracy of weather forecasting, for example, will require increased computing. Further automation of air traffic control to handle increased traffic in all weather conditions will require more complex data processing. New applications, not yet conceived, may be expected to place new demands on data processing power.

New larger systems will continue to be developed, because automation of information processing increases man's capability to utilize information. Each advance in information processing capability increases the number of ways in which it is possible to do things—to operate a business, to affect the economy, or to defend against enemy attacks. Some of these ways are better than old ways; they cost less, they get things done sooner, or they make it possible to do something that could not have been done before—like sending a man to the moon, which could not have been done without automated information processing.

Since automated information processing requires programs to direct the sequence of computer operations, the size and complexity of an information system is manifest not only in the power of the computing hardware but also in the software—the collection of programs required to make the system function. In the largest operational systems, the software may contain millions of instructions distributed in tens of thousands of modules. New larger systems may then be expected to be in the range of ten to one hundred million instructions and to be composed of several hundred thousand modules. The complexity of this software is due not merely to the large number of instructions and modules but to their logical interactions. Development and maintenance of such systems will require significant advance in the tools and techniques supporting software developers and maintainers.

Tool Integration

Early in the history of computer based information systems, tools were developed to support software development. A software tool is a program that automates a portion of the software development process, generally performing some tedious chore. For the most part, each tool is used individually and requires some set up—i.e., connecting the tool to the program and supplying the data necessary to specify precisely what the tool is required to do. Additionally, many tools were specialized to a particular project (because it cost less to do so) and are not directly usable on another project, necessitating development of essentially the same tool (with project specific variations) over and over again. Recently attention has turned to developing "programming environments"—integrated collections of software tools, with each tool in the collection usable in a wide variety of situations.

Examples of software tools are

* *Assembly languages* and their associated assemblers, which automate assignment of storage locations and assembly of subroutines.
* *Higher order languages* (e.g., FORTRAN, COBOL, Ada) and their associated compilers, which provide powerful expressions, reducing the number of statements the programmer must write, and which check the syntax of programs.
* *Editors,* which support on-line, interactive writing of programs.
* *Debuggers,* which monitor program execution, supporting the detection of errors by displaying values of specified variables at checkpoints.
* *Test tools,* which instrument programs so as to record during test execution the code segments exercised and the branches taken [4].
* *Programming design languages,* which provide a format for program design and check design adherence to that format [5, 6].
* *Requirement specification languages,* which provide a precise language for specifying software requirements and which check requirements written in the language for consistency and completeness [7].
* *Requirement traceability tool,* which supports tracing of requirements to design elements to code.
* *Word processors,* which support the preparation of program documentation.

The UNIX* operating system [8], developed by Bell Telephone Laboratories, took an important step toward tool integration by providing a pipe feature, which "pipes" the output of execution of one program to

---

* UNIX is a trademark registered by the Bell Telephone Laboratories.

another program where it is used as input to execution of that program. This and other UNIX features stimulated the development of UNIX based software tools, making UNIX a popular system for software development.

In its plans for Ada**, the new standard programming language for mission critical software development [9, 10], the Department of Defense recognized, early in the language development, the need for tools to support Ada software development and for tool integration. These needs were expressed in the Stoneman [11] requirements for Ada Programming Support Environments (APSEs). Two minimal APSEs are under development, one by the Army Communications Electronics Command, the Ada Language System (ALS), and the other by Rome Air Development Center, the Ada Integrated Environment (AIE). Both these APSEs will provide an Ada compiler, a Kernelized Ada Programming Support Environment (KAPSE), a linker and loader, and a minimal set of tools supporting software development. The KAPSE includes the Ada run-time system and a data base supporting tool integration and software development. It contains the principal machine dependencies of the APSE. Tools access machine facilities and interact with each other through the KAPSE.

To provide for and to encourage the development and integration of additional APSE tools, the Department of Defense formed the KAPSE Interface Team (KIT) and the KAPSE Interface Team, Industry and Academia (KITIA) to develop standards for tool interoperability and transportability. The KIT is developing Standard Interface Specifications (SIS) which APSE tools and the KAPSE are to be required to meet. A principal objective of the SIS is to ensure that any tool conforming to the SIS can be transported without change between computer systems having conforming KAPSEs and that the transported tool will be interoperable with other tools in the new system.

Tool integration automates the transition from one tool to another, passing data from the first tool to the second and retaining the context of the tool application. For example, a debugger tool supports debugging of a program by inserting breakpoints at which program execution will stop and the values of specified variables will be displayed. After the programmer has observed the values of the variables at a breakpoint and determined that a problem exists, he may wish to examine a section of the code and change it to correct the problem. For this the programmer would call another tool, an editor; however, in making the code changes, he may need to reference the variable values used in identifying the

** Ada is a trademark registered by the Department of Defense.

problem. When the changes have been made, he would like to execute the program again for the same test case and observe the variable values at the same breakpoint. If the debugger and editor are separated, non-integrated tools, calling the editor would lose the values of the variables displayed at the breakpoint, and the program to be edited would need to be identified to the editor. When the edited program is to be tested, the debugger would have to be called, the data values for the test case would have to be entered, and the breakpoints would have to be specified again. Making the transition from debugger to editor to debugger can be time consuming and subject to errors in the transcription of data and re-entering of data. For other combinations of tools, the transition can be even more time consuming and require entry of more data. Tool integration simplifies the process by automating the transition and preserving the output and context of the application of the first tool, improving programmer productivity by speeding up the process and reducing transcription and data entry errors.

The first attempts at tool integration involved designing two tools to communicate with each other. This produced some benefits but, as the number of tools to be integrated increased, this approach presented some problems. As each new tool was added to the tool set, some of the other tools had to be changed to work with the new tool. Because tools communicate with each other by exchanging data, the approach of tool integration through a data base evolved. This is the approach taken in the Ada Programming Support Environments. The data entered to use a tool and the data produced by the tool are entered into the data base. A command language is used to invoke the tools and to reference the data base.

At TRW, an important step in software tool integration was taken with the development of the Software Productivity Project (SPP) system [12]. SPP uses the University of California, Berkeley, UNIX system on the VAX 11/780 computer. To the UNIX tools, several tools developed at TRW were added. Among these tools were: Requirements Traceability Tool (RTT), Software Requirements Engineering Methodology (SREM), and FORTRAN Analyzer. Tool integration was accomplished through a unified data base implemented using the UNIX file system and the INGRES relational data management system. The relational data base contains relations on text files (program text, documentation, etc.) stored in the UNIX file systems. To improve performance, the relational data base is now stored on the Britton-Lee IDM 500 Data Base Machine. Tool integration is proceeding in an evolutionary manner as SPP users and system designers learn how to use tools in a coordinated fashion on software development.

The computer system on which SPP is implemented is shown in block diagram form in Figure 14.1. It shows users interacting with SPP through personal terminals connected through bus interface units (BIUs) attached to a local network bus. A gateway computer connects it to other local networks containing other computers. SPP software structure is shown in Figure 14.2.

During the approximately year and one-half the SPP has been in operation, it has established the value of tool integration in terms of increased programmer productivity. Although objective measures of productivity improvement are difficult to obtain, subjective evaluation estimates the increase at over 30%.

The increased automation of software development provided by tool integration will contribute much to the development of large systems. In addition to the productivity increase, tool integration makes large system development economically feasible, for the capability to use several tools in a coordinated fashion significantly aids in the management of

PRODUCTIVITY SYSTEM: HARDWARE OVERVIEW

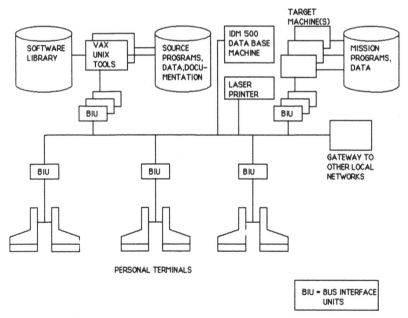

Figure 14.1. Productivity system: hardware overview

PRODUCTIVITY SYSTEM: SOFTWARE OVERVIEW

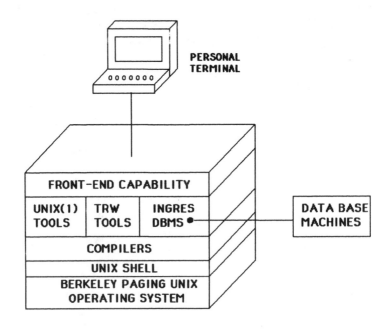

(1) UNIX IS A TRADE MARK OF BELL LABORATORIES

Figure 14.2. Productivity system: software overview

the complex relationships present in large systems. For example, effective use of multiple tools can determine the effect of code changes in one software module on other software modules and on system design and requirements documents, making up-to-date system documentation possible.

Formal Model of Software Systems

A formal model of software systems [13] provides a framework for developing the coherent picture needed for effective management of large software system development. Specifically, the model provides a formal representation of the software system structure, software requirements, the software design, and software testing. It maps the requirements to the design, implementation, and test cases. By providing this explicit representation of system development relations, the model reduces the apparent complexity of large software system development. Yet the model is relatively simple, formalizing one's intuitive notion of a software system as a collection of programs organized to perform specified functions.

Model of Software Structure

The model is based on the definition of a program given in the SEMANOL system [14]: A program p specifies the computation of a function f on the set E of possible inputs to p. E, the input domain of p, is composed of members E_i, the set of input values required for an execution of p.

$$E = \{E_i : i = 1, 2, ..., N\}$$

Each E_i is a set of ordered pairs, $E_i = \{(v_1, a_1), (v_2, a_2), ..., (v_n, a_n)\}$, associating input variables v_r with input values a_r. The input values a_r for an E_i include all values necessary for an execution of p, including any values saved from a previous execution of p or in a data base accessible to p—i.e., state variables affecting the execution of p. The function f is a rule assigning to each E_i a value $f(E_i)$ from a set called the range of f.

The model partitions the input domain E into subsets G_j corresponding to the inputs causing execution of logic path L_j. Since the code sequence of logic path L_j is a program, L_j specifies the computation of a function f_j defined on G_j. The total function f on E specified by p is therefore represented as a collection of functions f_j on G_j. The subsets G_j are disjoint and collectively cover the set E. Figure 14.3 shows input domain partitions and how a logic path L_j connects a value from an input domain partition to an output value.

Since the model associates the functions f_j performed by the program with logic paths L_j, it suggests that the number of logic paths may not be as large as the enormous numbers frequently computed from the combinatorics of branch expressions and used to convey the impression of large programs having unanalyzable complexity. Analysis of actual programs [15] has shown that the number of executable logic paths is a small fraction of the number of apparent logic paths computed from branch expression combinatorics. The unexecutable logic paths arise when branch expression evaluations are incompatible with assignments of values to variables appearing in a branch expression; e.g., the branch expression X.LE.1 in a FORTRAN IF statement would be evaluated FALSE in a program containing the assignment statement $X = 2$ prior to the occurrence of the IF statement. The existence of unexecutable logic paths in most programs is an artefact of commonly used programming techniques, for such programs can be rewritten [15] as "functional"

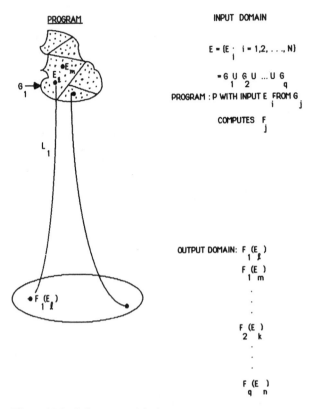

PROGRAM

INPUT DOMAIN

$E = \{E_i \cdot i = 1,2, \ldots, N\}$

$= G_1 \cup G_2 \cup \ldots \cup G_q$

PROGRAM : P WITH INPUT E_i FROM G_j

COMPUTES F_j

OUTPUT DOMAIN: $F_1(E_\ell)$
$F_1(E_m)$
$\cdot$
$\cdot$
$\cdot$
$F_2(E_k)$
$\cdot$
$\cdot$
$\cdot$
$F_q(E_n)$

Figure 14.3. Software model elements

programs containing no unexecutable paths. The general occurrence of unexecutable paths has contributed to the illusion of near infinite complexity of computer software, although programs for real world problems must compute only a reasonable number of functions.

Software Requirements

Corresponding to the functions f_j and G_j a program actually computes are the functions $\hat{f}_j$ and $\hat{G}_j$ the program is intended (required) to compute. The f_j and G_j may differ from the $\hat{f}_j$ and $\hat{G}_j$ owing to errors in implementation. Thus the model represents functional requirements as well as program structure.

At the user requirements level, the number of functional requirements is generally relatively limited. Additionally, individual requirements proposed by prospective users of a program may be stated as relations $\hat{R}_j$ rather than functions, allowing the output value computed for a given input to be chosen from a (usually limited) set of values rather than the single value required by the function—e.g., specifying the output variable y to be the range $a < y < b$. In this case, the implementor must complete the requirements, selecting the function $\hat{f}$ to be implemented by the program p. Parnas [16] has investigated the case of incomplete requirements in terms of "limited domain relations" (R_j, G_j) corresponding to non-deterministic programs, which allow environmental conditions to select the function to be computed.

The functional requirement representation in the model is compatible with the format prescribed for B-5 software requirement specifications in MIL-STD-490 [17], viz., in the B-5 specification, inputs correspond to $\hat{G}_j$, processing corresponds to $\hat{f}_j$, and outputs correspond to $\hat{f}_j(E_i)$. However, analysis of B-5 specifications using the model [18] has shown they generally are incomplete. The list of input variables in the input section of the specification tends to be incomplete, and generally the ranges of the variables are not specified; the processing section may fail to define functions for all input domain partitions; and the list of output variables in the output section also tends to be incomplete. This defers part of requirements definition into design and coding phases, where requirements misinterpretation is apt to occur as designs and code are developed from incomplete requirement specifications. Using the model as a guide to writing complete software requirement specifications takes more time but results in fewer design and implementation problems and shortens the design and coding phases.

In addition to functional requirements, a software system generally has other types of requirements, such as performance requirements or security requirements. These types of requirements may be included in the model in terms of constraints Q_j on the functions f_j.

Software Design
Software design transforms functional requirements into implementable functions organized into suitable modules. Although in some cases the functional requirements defined by system users may be directly implementable functions, this may not be so in general, particularly for large systems. The requirements may be global in nature and need to be decomposed into smaller pieces to be readily implementable. The model of software structure and software requirements can be extended to represent the decomposition of functional requirements into functional design elements.

In most large projects, the software design is developed in two steps: a top-level design and a detailed design. The top level design defines the external interfaces of the software and its major structural entities. For computer operating systems, the top level design defines the system calls that application programs use to call operating system services. In this case, the model represents each operating system service as a function $\hat{f}_j$. The $\hat{G}_j$ on which $\hat{f}_j$ is defined specifies the parameters involved in the system call and any state variables affecting the service. The detailed design details the top level design into smaller design elements, carrying the requirement decomposition down to the requirements on each sub-routine. Then the design elements will correspond directly to the software structural elements.

Decomposition of a functional requirement $\hat{f}_j$ on $\hat{G}_j$ into the "smaller" functional requirements needed in a detailed design involves two kinds of decomposition processes:

- *Domain partitioning*, partitioning the input domain partition $\hat{G}_j$ into smaller partitions $\hat{G}_{ji}$ on which functions $\hat{f}_{ji}$ are defined, and
- *Functional decomposition*, decomposing the function $\hat{f}_j$ into a sequence of functions—e.g., $\hat{f}_j = \hat{b}_j \hat{C}_j$ in which the function $\hat{C}_j$ is evaluated first and $\hat{b}_j$ is evaluated at the point defined by the value $\hat{C}_j(E_i)$.

A detailed design generally will require both types of decomposition. When all of the requirements are decomposed into detailed functional design elements, some of the functions may turn out to be the same; i.e., the same rule assigning output values to input values is applied to different input domain partitions. These two input domain partitions may be combined into a larger partition on which the function is applied. Eliminating such functional redundancy can simplify the design and has been found to improve performance.

A central point of using the model is maintaining the discipline of traceability of requirements to top level design elements to detailed design elements to software structural elements. Many of the problems of large system development have been due to failure to maintain this traceability. Then the top level design elements may not correspond to functional requirements; detailed design elements will generally not correspond to the top level design; and the code will not correspond to the detailed design. Thus no coherent picture of the system is available, and verification that the system satisfies the requirements is extremely difficult and costly. To maintain this traceability among the large number of elements in large systems, a tool such as the Requirement Traceability Tool mentioned before is essential.

Testing

The purpose of testing is to demonstrate satisfaction of requirements and to find problems before the system is placed in operational use. The model, through its functional representation, relates test cases to requirements. Each test case may be characterized by the input E_t initiating its execution. That test case is a member of an input domain partition G_j. By choosing test inputs from the $\hat{G}_j$ of a functional requirement, test cases for demonstrating satisfaction of that requirement are obtained. Thus if the $\hat{G}_j$ of the functional requirements are defined in the requirement specification, test cases to demonstrate satisfaction of requirements can be defined at that time, and each test case is associated with the requirement whose satisfaction it is to demonstrate. Choosing test cases in this way has been shown to provide thorough demonstration of requirement satisfaction with substantially fewer test cases than when they are chosen by conventional methods.

The model contributes to the development of large software systems by providing a formal description of the system in terms of the functions the system is required to perform, the decomposition of the functions into design elements, the correspondence of design elements to software structural elements (routines), and test cases for demonstrating requirement satisfaction. Thus it provides a coherent picture of the system, its development elements, and system documentation, defining relationships among them. Applied to new distributed computer architectures, the model aids the analysis of a system into functions assignable to individual computers.

References

[1] Metropolis, N., and Nelson, E.C., (1982) Early Computing at Los Alamos, *Ann. Hist. Comp.* **4**(4).

[2] Distaso, J.R., Kohl, B.F., Krause, K.W., Shively, J.W., Stuckle, E.D., Ulmer, J.T., and Valembois, L.R., *Developing Large Scale Real-Time Software: The BMD-STP Experience*, TRW Software Series 79-01.

[3] Williams, R.D., *Managing the Development of Reliable Software, Proceedings of the 1975 International Conference of Reliable Software*.

[4] Brown, J.R., *Practical Applications of Automated Software Tools, Proceedings of 1972 WESCON*.

[5] *PDL/74 Program Design Language Reference Guide* (1977) Caine, Farber, and Gordon, Inc., Processor Version 3.

[6] Sammet, J., Waugh, D.W., and Reiter, R.W., Jr., *PDL/Ada Design Language Based on Ada, ACM 81 Conference Proceedings*.

[7] Alford, M.W., *Software Requirements Engineering Methodology (SREM) at the Age of Two, Proceedings of IEEE COMSAC 78*.

[8] *UNIX Programmer's Manual* (1979) Seventh Edition.

[9] *ANSI/MIL-STD-1815A* (1983) Ada Programming Language.

[10] DOD Directive 5000.31.

[11] Buxton, J., (1980) *Requirements for Ada Programming Support Environments: Stoneman, U.S. Department of Defense.*

[12] Boehm, B.W., Elwell, J.F., Pyster, A.B., Stuckle, E.D., and Williams, R.D., *The TRW Software Productivity System, Proceedings of the 1982 International Software Engineering Conference.*

[13] Brown, J.R., and Nelson, E.C., (1978) *Functional Programming, Final Technical Report*, RADC-TR-78-24.

[14] Blum, E.K., (1970) *The Semantics of Programming Languages, Proceedings IFIP Working Group 2.2.*

[15] Nelson, E., (1981) Functional Programming Analysis, *J. Syst. and Software* 2.

[16] Parnas, D., (1981) *An Alternative Control Structure and Its Formal Definition, IBM Technical Report* TR FSD-81-0112.

[17] *MIL-STD-490, Military Standard Specification Practices*, U.S. Department of Defense.

[18] Logan, J., (1980) *Software Reliability, Application of a Reliability Model to Requirements Error Analysis, Proceedings of the Fifth Annual Software Engineering Workshop.*

15   Reflections on Style in Physics

J.H. Manley

The topic I have chosen, style, is not subject-matter physics; nor is it a subject on which I claim recognition. My discussion will not put forth carefully reasoned conclusions from laboratory data as would be the case for a paper asking to be accepted in the society of scientific facts. Its purpose is more to set forth some reflections to stimulate thought about this matter of style. Although I speak of physics in my title as a reminder of my own background, most of my discussion is also appropriate to the broader term, science, which I shall use frequently.

As a topic, style of doing or practicing physics has some advantages. It can be treated by temporal comparison and thus avoid traps of absolutism. In this manner a rich collection of memories, though not necessarily reliable as future predictors, may be used as background to discuss changes in style without being obtrusively reminiscent. As I plan to explain, I believe my topic is timely, for circumstances today demand from all scientists more attention to and participation in problems beyond a discipline and a specialty. This timeliness is suitable as a bequest to a generation younger than mine.

My topic also has disadvantages. An obvious one is that from my somewhat disengaged, limited, and senior position I cannot be fully and intimately acquainted with practices, people, and character of present

NEW DIRECTIONS IN PHYSICS
The Los Alamos 40th Anniversary Volume
ISBN 0-12-492155-8

day physics. The chief disadvantage, however, is the difficulty of imparting exactly what one means by "style". I remember a frequent and favorable assertion, even an emphasis, of Robert Oppenheimer on style, but I confess I never knew quite what he had in mind. This assertion was in the same class but less obvious than another expression of his, "technically sweet". This expression could describe something in physics or technology as approbation of the thing itself, its style, quite unassociated with any individual or personal style. Even the term "personal style" has uncertain connotations. An individual has a style quite easily observed by an acquaintance but a physicist may refer to something somewhat different, his style of doing physics. My own recollections tend to emphasize first a style of doing physics, but some examples will illustrate that two aspects of personal style are intermixed. I think of a Fermi style deriving from a deep and extensive command of mathematical techniques, a breadth of knowledge of physics so fruitful for originality and versatility as attention is transferred from one area to another. All these elements of style at Fermi's command were crowned by his ever-practical approach to problems to result in a clarity and simplicity that delighted us all. Oppenheimer had a social style of leadership and of doing physics. I have a number of memories of him and a colleague pacing up and down some railroad platform engaged in intense talk. These two are close to conveying a style of doing physics. Gamow, unique in stature, had a style best seen in informal English which was such a marvel of originality in spelling and syntax that it became known as a different language—Gamowian. Teller has many unusual elements of style, but I content myself by mentioning only his bushy eyebrows and their assistance to his pontifical manner of speaking. Bohr, it seemed to me, conveyed the profundity of his thoughts by speaking so softly as to be scarcely audible to the listener. In my reflections, the styles of Gamow, Teller, and Bohr are connected with personal eccentricities. However interesting a discussion of personal style in this manner might be from a gossipy, earthy viewpoint, another phenomenon of style or, more accurately, of its changeability intrigues me.

In my lifetime I have seen a major departure from earlier norms of style in the profession of physics. This departure or change has taken place or at least been initiated in a short time compared to past evolutionary rates of change. Because of this, one could call this change a revolution in style. Although like many other changes it has root causes in population growth and the pressures thereby placed on individuals, its abruptness must have been caused by additional influences. I think these are the cataclysmic events we designate as World War II. However, to reach an understanding of the significance of the meaning of this change, I must examine forces which create changes in style.

One attribute of style is a strange duality. It is individual yet collective. Individuals found schools in art, literature, music, even dress, which become collective manifestations of the individuality of the founder. This kind of personality extension then becomes a force which dictates what is acceptable—a form of peer pressure.

In science, too, the interaction between the individual and his professional group causes changes of style in each. This interaction has been discussed by Thomas Kuhn in his book on the structure of scientific revolutions [1]. In this book he develops his paradigm idea and credits a 1935 paper by Ludwig Fleck as anticipatory to his own analysis [2]. Fleck's relatively unknown paper would startle many of us today, for he advances the thought that scientific *facts* are *made* things. This idea grows from his argument that scientific thought is holistic in character, deriving from individual "thought style" (Denkstil) and a group "thought collective" (Denkkollektiv). I would not easily concur with his conclusion about scientific *fact,* but I would for *style.* As I see it, it is just this interplay between individual and group thought which continually modifies the style of each. The style of the individual is collectivized and altered to that of the group. This in turn feeds back to cause individual adaptation. In this view, to speak of "doing physics with style" may only mean proceeding in a socially acceptable manner, possibly only to the elite physicists—those whose physics may not depart greatly from the norm yet contains sufficient originality or discovery to bring recognition.

I can elaborate by quoting I.I. Rabi who became my first post-Ph.D. mentor when I went to Columbia 52 years ago. Rabi went to Europe not to learn the literature of physics, which he knew, but to "learn something which I find very difficult to describe" [3]. He explains by saying, "I knew the score but not the melody." Rabi went abroad to study with an individual who had style so that he himself could do physics with style. I say that Rabi's melody was not a simple single tune of an individual but the integration or collectivization of all tunes which sound pleasing and harmonious to physicists. At that time European science expressed the accepted collective style. Study with one individual usually was an adequate exposure. An example of operation of collective style is the attraction of students to an opportunity to study under a Nobel Laureate. It is a statistical fact that such students themselves receive more Nobel Prizes than do students of non-Laureates. Evidently the style learned has group approval.

A determinant of style is any factor which limits it. Art and artists are described by attributing a certain style thereto. I think the final impression of the viewer is limited by the physical properties used by the creator, by the image the creator has in mind, and by the artistic conditioning and experience of the viewer. The creation of a world picture by a physicist has similar limitations and an additional major one: The picture must not do violence to any facts that are observable, repeatable, and synthesizable by any competent colleague. This limitation, arising because Nature is asked by experiment to reveal her style, is a rigid one. The rigidity has its origin in the constancy of the truths which supply the fabric for the world picture. Other determinants of style provide limits that change and evolve. These are usually those imposed by human elements of one kind or another. An example of such is found in the writings of Robert K. Merton. Looking at early days of the Royal Society, he sees stimuli and guidelines of its founders in the Puritan values of diligence, rationality, asceticism, self-denial, civic spirit, and service to God through works. In this example the strength of the limitation the guides impose has certainly diminished with time. Possibly one of the most persistent is a kind of selflessness which is still acceptable even though in apparent conflict in an individual with his desire for recognition and status.

Some other determinants of individual and group style do not so much set limits as act as agents of change. Examples are growth in number of physicists, financial support, and social pressures.

It was not unusual in my early professional days to be acquainted with most physicists no matter what their specialty. Today one can know only a fraction of those even in his own field. With this growth in the number of scientists is that of journals and papers, a flood which requires more and more time and makes great difficulty in keeping familiar with the work of others. The flood would be even greater had not multiple authorship, arising from observational complexity, become a style. Multiple authorship is also symptomatic of another style change, the decline of the generalist and the growth of specialists. Part of this growth must be associated with the general increase of population—more students in the educational mill, more utilization of science in expanding industrial activity (especially new high-technology enterprises). Another part stems from markedly increased financial support.

Much of the growth of support of science has occurred in the last forty years and comes from Federal funds, generated by concerns of national defense and scientific status. It has taken the form of creation of new national laboratories and widespread contractual support of industry and academia. For the physicist it has some features of paradise. Even a small laboratory can now buy equipment instead of making it. Big laboratories with huge accelerators make possible far more complex experiments directed toward seeking answers to even deeper questions about matter than before. As apparatus and equipment become more complicated and sophisticated than in the sealing wax and string era, so data reduction passes beyond the slide rule and hand computer to modern computer technology. Computers add to their usefulness by becoming instruments to accomplish experiments themselves. Their inputs can be questions as well as observations.

These changes in growth in numbers and in funding have a profound effect on style. There has been increased peer pressure in the professional community, pressure to attain status, and pressure to obtain satisfying employment. These and similar pressures are very determinative of individual and collective style. The result is a change in the attitude of physicists toward one another. Each feels compelled to operate with a new concern for his own welfare. The Puritanical good of civic spirit appears replaced by a more selfish and competitive game. The game involves the generation of funding proposals to obtain grants and periods of professional or public service on various local, state, and national boards or committees. This public service may be undertaken more for the exposure it provides than for inner satisfaction of performed civic service. An associated activity, public appearances, may be motivated more for desire for prestige and power than for the education of one's fellows. Some scientists spend considerable effort in experimentation with promotional and managerial techniques. In some of these activities there may be occasional distortion of older ethical standards. As someone has said, a young physicist today cannot afford the luxury of asking, "Is that right?" He asks instead, "Why did he say that?" When expediency replaces principle as a personal guide, estrangement grows between one individual and another. We cannot be sure of a colleague's statements or retain confidence in a long-term pattern of individual and professional relations. We may hope that integrity persists in published scientific works, but there are disturbing reports of outright data falsification in some sciences—though not in physics as far as I am aware. A recent opinion piece in the science magazine, *Nature*, entitled "Is Science Really a Pack of Lies?" reviews the alarm in the community as a result of a spate of investigations of fraud.

An adjunct effect of pressure on individuals could extend to the international community. Traditionally science has been international in character and fundamentally so. As such it has contributed favorably to relations between nations through a bridge of individuals bound by common interest in universal phenomena. If the estrangement between individuals of which I speak spreads as a contagion to national groups, it could create a reversal of this positive influence and assist in a new tragedy of isolation.

Discussion of the third determinant of style, social pressures, is more involved than the other two. Some of the pressures already treated are social pressures within the professional group. However, both the individual and this group are part of a larger organization, society. Another characteristic of today's world is the increasing flow of style-modifying currents between society and individuals and between it and professional groups.

There was a time when scientific interests were supported by private income, patronage of the rich, or perhaps a few institutions rather insulated from the work-a-day world. Then the coupling between individual and group was weak and style very individualistic. With expansion in interest and in numbers the individual-group interaction strengthened and so did the forces for style modification. However, it was still an ivory-tower period which fostered the belief, especially among scientists, that science and society had little in common. Scientific discoveries and technological derivatives were indeed taken into society at large, but this was no concern of the scientist. He and his fellows exerted some influence on society by the style-bearing currents flowing thereto, but the reverse flow was sufficiently small that the independent, ivory-tower position persisted. Science, like the arts, was considered a cultural subject, proper for some educational programs but largely disregarded by labor and business.

It was clear that if stronger inter-relations between science and society were to develop, another style-modifying interaction would become operative. Style would be subject not just to the individual-professional group currents, but with the inclusion of the larger group, society, there would be additional flows between pairs: individual-society and professional group-society. These currents, proportional to the respective interaction strength, would all exert their effect on style in the scientific disciplines. In terms of the ideas of Fleck, relating the individual "thought style" and the professional group "thought collective" there must be added another "thought collective", that of society.

I found it interesting to discover recently a parallel to these ideas in economic theory. In J.K. Galbraith's *The New Industrial State*, Chapter XIV, his "principle of consistency" states that there must be consistency in the goals of society, the organization (industrial in his case) and the individual. Obviously, style is ultimately an expression of agreed-upon procedure to reach goals.

Upheavals in society initiate somewhat abrupt but persistent change in the magnitude of style-changing currents between society, the professional and his group. The industrial revolution was one such upheaval. Two wars, WWI and WWII, designated in folklore fashion as a chemists' war and a physicists' war respectively, are others. All such events, which have initiated the realization of the importance of science and technology to society, have produced major changes in style for scientists and their professional groups.

The last forty years have seen a tremendous change in the interaction between science and society. This was certainly initiated by World War II and then maintained and augmented by requirements of national defense and by the rapid growth of civilian demand for products of science and technology. My earlier reference to growth and Federal support as causes of style change within the physics profession can now be seen as part of the result of multiple interactions between individual, professional group, and society. Developments such as those in nuclear energy, materials, communications, transportation, high technology, and so on created a social demand for more science and technology and a favorable atmosphere for appropriations from the public purse.

The flow of materials and money between science and society has quite understandably been accompanied by a less tangible but very real flow of scientific and technical data and advice to units of government. Scientists are very aware of this kind of flow to the Federal government. Part of its substance expresses a very natural concern over how Federal science policy will affect their work. As an example of this type of flow, I have in mind the lobbying efforts of a young group of scientists in 1945–46 to secure congressional passage of a bill which would terminate military administration of the nation's atomic energy activity and place it under civilian direction. It was a spontaneous, remarkable demonstration of the emergence from ivory towers to public halls which initiated a new era of science-government relations. By now we have become quite accustomed to scientists appearing before Congressional committees and being members of various federal boards, as well as being Presidential Advisors and guests of information media for public exposure of their views (not always wholly scientific) and so on. All this has brought about a major upheaval in style.

In this type of public activity scientists are in a difficult position. Their customary style emphasizes rationality, logic, and established factual information. Superb as this style has proved itself to be in understanding phenomena involving a limited number of variables and devoid of human unpredictability, it is not ideally suited to the solution of human problems. With a modicum of attention scientists might realize that if they express an opinion that something is technically sweet in their restricted meaning of being logically consistent or an admirable application of the laws and techniques of science, their audience may quickly but incorrectly conclude that this development is approved for human use. It is as if a new drug, fascinating in structure and beautifully consistent with the rules of organic chemistry, is thereby automatically approved for consumer consumption.

Some scientists have conveyed the impression that any problem, given application of enough time, money, and professional expertise is soluble by that offspring of science, technology. This prescription of a "technical fix" for individual human problems or large collective ones such as war and peace should carry a label, "DO NOT NECESSARILY BELIEVE. MAY BE VERY DANGEROUS TO YOUR HEALTH!" The goal of a technical fix furnishes livelihood for many of us; it pays for a great deal of the fun we have with technically sweet research and our first love, basic research. But in the process of pursuit of such a goal, our style, like our theoretical and experimental topics, may be unduly influenced by "where the money is". The dilemma is to achieve proper balance between our roles as scientists and as citizens.

I do not know a proper or unique path through this forest of often incompatible interests. It is much easier to "view with alarm" as I have done in these comments about interactions and style. In our professional life I hope that reflections such as these may increase our awareness of our style and the forces which tend to change it. I am sure that there is no returning to the ivory-tower style of the past. That is as it should be. The forces to which I have assigned responsibility for style-change will continue. Population growth cannot be expected to alter appreciably for generations. There is no apparent reason for any lessening of society's demand for the fruits of science and technology, and demand is accompanied by societal, financial, and other support. Many nations aspire to equality with the so-called developed countries and our nation continues to feel the influence of strengthening interactions with other powers both large and small and the accompanying effect on our own style. World problems now and in the future require conscious application of the best minds for their solution, and scientists have a problem-solving experience to link with that of non-scientists. It might just happen, too, that some of their early professional style derived from high principles could help in the tasks ahead.

References

[1] Kuhn, Thomas S. (1962; 1970). *The Structure of Scientific Revolutions*. 2nd edition, enlarged, Chicago, pp. vi–vii.

[2] Fleck, Ludwik (1935; 1979). *Genesis and Development of a Scientific Fact*. English edition, Chicago. First edition, Basel, Switzerland: Schwabe & Co.

[3] Rabi, I.I. (1987). "How Well We Meant" in *New Directions in Physics: The Los Alamos 40th Anniversary Volume*. Boston: Academic Press, pp. 257–265.

16   Tuning Up the TPC

Owen Chamberlain

The name TPC stands for Time Projection Chamber. It is a new style of particle detector invented in Berkeley by David Nygren. It is the core particle detector in an apparatus called PEP-4 designed to detect outgoing particles produced by electron-positron collisions at the accelerator called PEP.

The PEP accelerator at the Stanford Linear Accelerator Center (SLAC) is a colliding-beam machine (an electron synchrotron) where electrons and positrons collide head-on. Each beam has an energy, typically, of 15 GeV, giving 30 GeV in the center-of-mass system.

The TPC is built around the vacuum pipe in which the two beams circulate and interact. The total PEP-4 detector and its TPC core are the product of a large collaboration of some 50 scientists[1] from six in-

---

[1] Because I am only one of these many scientists it is important that I list the people involved: H. Aihara, M. Alston-Garnjost, D.H. Badtke, J.A. Bakken, A. Barbaro-Galtieri, A.V. Barnes, B.A. Barnett, B. Blumenfeld, A. Bross, C.D. Buchanan, W.C. Carithers, O. Chamberlain, J. Chiba, C.-Y. Chien, A.R. Clark, O.I. Dahl, C.T. Day, P. Delpierre, K.A. Derby, P.H. Eberhard, D.L. Fancher, H. Fujii, B. Gabioud, J.W. Gary, W. Gorn, N.J. Hadley, J.M. Hauptman, B. Heck, H.J. Hilke, W. Hofmann, J.E. Huth, J. Hylen, H. Iwasaki, T. Kamae, R.W. Kenney, L.T. Kerth, R. Koda, R.R. Kofler, K.K. Kwong, J.G. Layter, C.S. Lindsey, S.C. Loken, G.W. London, X.-Q. Lu, G.R. Lynch, L. Madansky,

NEW DIRECTIONS IN PHYSICS
The Los Alamos 40th Anniversary Volume
ISBN 0-12-492155-8

stitutions: Lawrence Berkeley Laboratory, Johns Hopkins University, University of California–Riverside, University of California–Los Angeles, Yale University, and University of Tokyo.

The special feature of the TPC as a charged-particle detector is that it attempts to identify each produced particle, not just by its electric charge and its momentum, but also by its particle type—electron, muon, pion, kaon, or proton. Furthermore, it attempts to identify particle type *directly* for each outgoing particle.

The new feature of the TPC that allows the possibility of direct observation of particle type is its capability of accurately measuring the specific ionization along a track (the trail of ionized atoms left behind by a charged particle passing through gas). To give good accuracy in the specific ionization (the number of ion pairs per centimeter along the track), the gas pressure is made high—8.5 atmospheres absolute pressure.

The name "Time Projection Chamber", while not perfect as a name characterizing the instrument, has some basis as a descriptive name. It refers to the fact that electrons, released by charged-particles that pass through the gas, are timed by clock circuitry as they drift in a uniform electric field toward the end planes of the cylindrical instrument, and their drift time is used as a measure of the drift distance. Thus the time of detection of a clump of electrons at the end plane reveals how far the electrons had to drift (in the z direction, parallel to the cylinder axis).

The electrons are detected by a multiwire proportional chamber with a gas gain of about 1000.

Figure 16.1 shows the basic features of the instrument. Electric fields and magnetic field are all parallel to the z axis (parallel to the beam directions). A track of released electrons drifts toward the endcap of the roughly cylindrical chamber. If the track is fairly straight (because it is due to a high-momentum particle that bends little in the TPC magnetic field) it will register on all 180 of the sense wires of an end sector. A series of pulse heights is recorded, one measurement every 100 nanoseconds. We speak of time "bins", each 100 nanoseconds wide. Clusters of adjacent time bins that have non-zero signals stored can be identified and summed over to obtain the integrated ionization on each of 180 sense wires. Fifteen of these wires have under them, not the usual grounded

R.J. Madaras, R. Majka, J. Mallet, P.S. Martin, K. Maruyama, J.N. Marx, J.A.J. Matthews, S.O. Melnikoff, W. Moses, P. Nemethy, D.R. Nygren, P.J. Oddone, D. Park, A. Pevsner, M. Pripstein, P.R. Robrish, M.T. Ronan, R.R. Ross, F.R. Rouse, R. Sauerwine, G. Shapiro, M.D. Shapiro, B.C. Shen, W.E. Slater, M.L. Stevenson, D.H. Stork, H.K. Ticho, N. Toge, M. Urban, G.J. Van Dalen, R. van Tyen, H. Videau, M. Wayne, W.A. Wenzel, R.F. vanDaalen Wetters, M. Yamauchi, M.E. Zeller, and W.-M. Zhang.

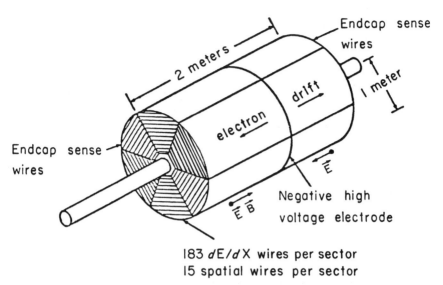

183 dE/dX wires per sector
15 spatial wires per sector

Figure 16.1. Basic structure of the TPC. The beam pipe runs through the center, along the z direction. The active volume is roughly one meter in radius and two meters long. The magnetic field is uniform throughout the TPC volume and in the z direction. The electric field is opposite in the two ends (halves) of the detector, causing electrons to drift toward the nearest endcap. Each endcap is divided into six end sectors. Each end sector has 183 proportional-counter wires, oriented as shown on the figure.

conducting plane, but a row of square electrodes that we call "pads". Each pad is attached to a charge-sensitive amplifier and gives a signal similar to the wire signals, but opposite in polarity, because the pad receives an induced signal rather than the direct signal.

Figure 16.2 shows the pad structure, with 8-mm × 8-mm pads. By center finding with a roughly gaussian spatial dependence of the pad signals the track coordinate at the sense wire is measured with an accuracy of 250 microns. This coordinate is important for getting good momentum resolution.

In describing the apparatus I shall put emphasis on the high-voltage insulation and the electrode structure intended to establish a very uniform electric field in the TPC, as that is the part I have worked on the most. My closest colleagues in this effort were Ronald Madaras, Alan Bross, and Bill Gorn.

In tests on a prototype insulator of about half final size we confirmed that high-value resistors, of perhaps 50 MΩ, could be damaged by the surges connected with electrical sparking. The manufacturer of our high-voltage resistors had warned us that damage could result if potential

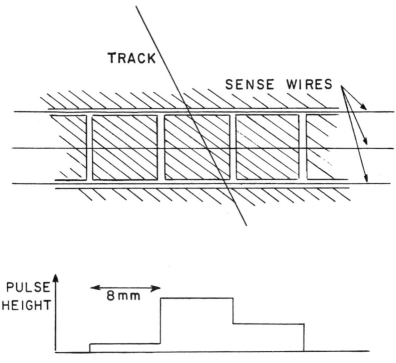

Figure 16.2. Pad row for determining the track position along a sense wire. The histogram represents the way in which the induced signal can be picked up by three pads.

differences of more than 4 kV appeared across one of these resistors. Therefore, we developed a "resistor package" in which a 2.2 kV spark gap is connected in parallel with each 50-MΩ resistor. In series with the resistor-spark-gap combination is a one-ohm "snubbing" resistor, intended to limit the current during a spark-down process, in which successive spark gaps fired, somewhat like a row of dominoes falling down in succession.

High-voltage tests also showed surface creepage phenomena along insulator surfaces, phenomena which could be prevented by interrupting the creepage path every 2 cm by a conducting line (made printed-circuit-board style).

On the basis of the prototype tests we developed a design for our high voltage system. In every case uniform potential gradients were to be established by long chains of precision resistors. Our resistor values were usually 50 megohms; a string contained 200 such resistors. They were manufactured to 1% accuracy, measured (each one) to 0.01% accuracy, selected to give matched foursomes of resistors to be connected in series. Each foursome had the same total resistance to within 0.01%. Within a foursome the order of the resistors was low-high-high-low. This gave an electric potential pattern correct to better than 0.01%. Accuracy of 0.01% would be maintained if the component resistors of each resistor chain suffered spontaneous random changes of no more than 0.1%. The uniform potential gradient was to be established by carefully laid out fiberglass (G-10) with a conducting line every 5 mm (center-to-center distance). The fiberglass and its conducting lines have been called a "field cage". We speak of a "fine field cage", whose elements define the electric potential at the boundary of the active volume. The fine field cage is surrounded by the "coarse field cage", which supplies an environment for the fine field cage in which the electric potential is correct to about one percent of the high voltage on the central (flat) membrane that is the high-voltage electrode. In typical operation our full high voltage is 75 kV.

The final instrument has unique features. It locates clumps of ionization unambiguously in three dimensions, with comparable accuracy in each of the three dimensions. In each dimension the accuracy is of the order of 250 μm. Track finding is spectacularly easy, particularly in a plot of r versus z, where there frequently are 100 or more points on one meter of track. The computer generally finds tracks easily and with high efficiency. Where the r-z plot reveals a clump of ions detected by a sense wire beneath which is a row of pads, the row of pads may be interrogated to determine an azimuthal position for the signal.

Furthermore, for most tracks there are about 180 pulse heights recorded; to be used for the calculation of specific ionization.

Figure 16.3 shows, on the left, the locations of detected ionization projected upon the x-y plane. In this figure only the pad information is used. Even so, track location is rather easy. On the right is a plot of r versus z. The tracks are easily followed even when tracks from several 60-degree sectors are plotted on top of each other.

Figure 16.4 shows the specific ionization versus momentum for several charged-particle types: electrons, pions, kaons, and protons. If the specific ionization can be measured with, say, 3% accuracy then many particles can be identified through simultaneous specific ionization-momentum measurement and, furthermore, particles can be identified at momenta above 1 or 2 GeV/c, momenta for which other methods such as the time-of-flight method are usually ineffective. (Other methods of identifying particles in this higher-momentum domain exist, of course (for example Cherenkov-counter methods), but they are seldom able to identify all of the outgoing particles in a complicated event.)

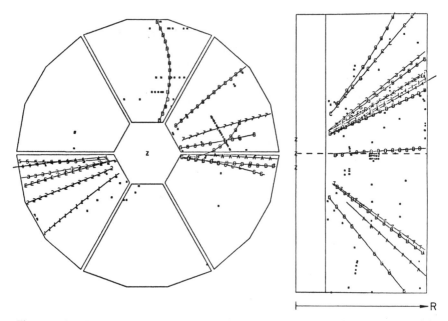

Figure 16.3. The one-event display; tracks reconstructed in a hadronic event. In the left portion of the figure the z axis (beam axis) runs perpendicular to the figure plane. The six sectors of one endcap are outlined. The right part of the figure is a plot of z, the coordinate in the beam-axis direction, plotted vertically versus radius r away from the beam axis plotted horizontally.

It happens that when 183 pulse heights are recorded, from 183 wires of an end sector of the TPC, the smaller pulse heights are more valuable than the larger pulse heights. This is because the pulse-height distribution that stems from an ionization distribution has a long tail on the large-pulse side of the peak, often called the "Landau tail". The larger pulses, being results of close collisions between the charged particle being detected and an electron in the gas, are subject to large fluctuations. To construct a measure of specific ionization, we discard the largest 40% of pulses and keep the lowest 60% of the pulses. These are averaged and that average is called the "truncated mean". It is usually our best indicator of specific ionization. Figure 16.5 shows the pulse height distribution of a single sense wire. Notice the long tail on the distribution for larger pulses. Figure 16.6 shows the experimental distribution of truncated means for a collection of outgoing particles, each of which has more than 120 wire signals from which to construct a truncated mean. These outgoing particles are chosen to be in the momentum region 450 MeV/c $< p <$ 750 MeV/c, for which the pion specific ionization stands at a

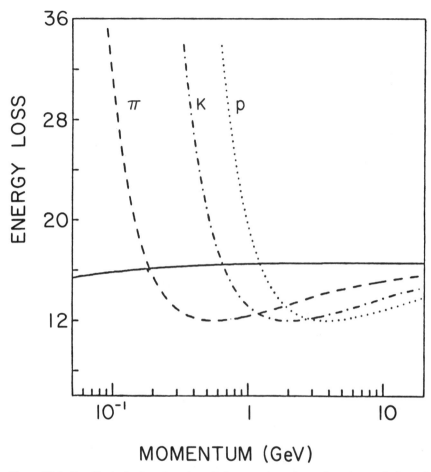

Figure 16.4. Specific ionization, loosely called energy loss, for various types of charged particle.

minimum well below the other curves in Figure 16.4. Thus Figure 16.6 shows a strong peak of pions with truncated mean pulse heights near 12 units and a small sprinkling of electrons, kaons, and protons between 15 and 20 units. For the pions the standard deviation of the truncated mean is now 3.5%, and we hope to reach a better figure of 3.3% when the analysis becomes more refined.

Figure 16.7 shows a scatter plot made from the data of hadronic final states produced by e^+e^- collisions. I believe the particle separation is in places quite persuasive, especially for momenta below 1 GeV/c. Above 1 GeV/c the results portrayed by the scatter plot seem to correspond

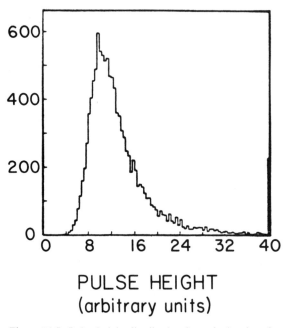

PULSE HEIGHT
(arbitrary units)

Figure 16.5. Pulse-height distribution for a single wire of a multiwire proportional counter.

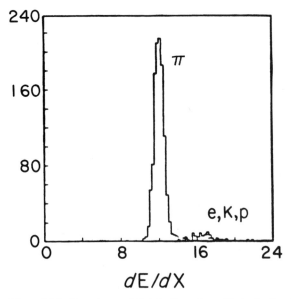

dE/dX

Figure 16.6. Experimental distribution of specific ionization based on the truncated mean.

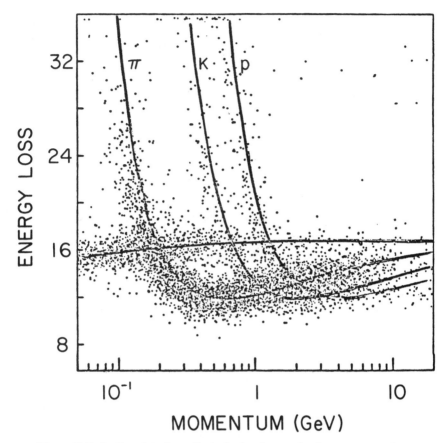

Figure 16.7. Scatter plot of specific ionization (energy loss) versus momentum.

well to our prior understanding of specific ionization. It is not unreasonable to argue that some points above 1 GeV/c can be assigned their particle type with some confidence. However, I don't think one can claim a clean *separation* of pions from kaons in the high momentum region.

Where the particle-type curves cross each other in Figure 16.7 there is definitely no way to escape an ambiguous result in the assignments of particle type. In principle one can detect a different width to the peak in the pulse-height distributions for different particle types. However, it does not seem likely that 183 samplings of the pulse-height distribution will be enough to yield much information in the "overlap" regions where the curves for different particles cross. We will in the near future be studying this question very carefully, nevertheless.

As you will see, I think, looking at Figure 16.8, the curves for each

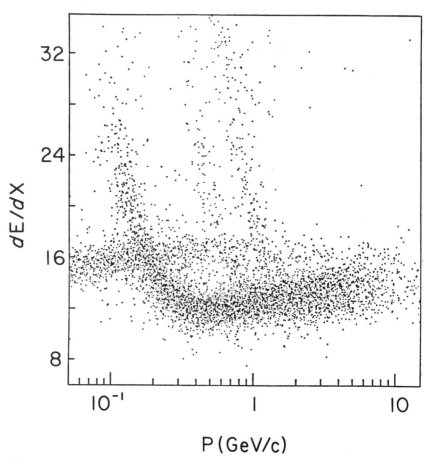

Figure 16.8. The same scatter plot as in the previous figure, this time shown without the curves for various particle types.

particle type are hardly necessary. One sees the lines quite clearly in some parts of the scatter plot.

Distortions

In cosmic ray tests of the TPC it was found that energetic muons that could be counted on to produce rather straight tracks were giving reconstructed tracks that had hook-like deviations near the edges of the active TPC volume. The nature of the distortion ("hooks") is that which you would see if electrons close to the field cages drifted more slowly than anticipated. Another feature of the results was that signals were typically missing from the three-or-so wires nearest the beam axis and missing from the three outermost wires. By mocking up the drift process on a computer, Raymond van Tyen was able to explain both the delayed arrival of electrons and the absence of signals on several wires near the edge of the active volume. His explanation involved electrical charges on the surface of the G-10 insulators of the fine field cage. We have accepted this explanation because we have looked for other explanations unsuccessfully; however, we are at a loss to explain why there are always *negative* charges on the surfaces near the endcap, which is the most positive part of the TPC structure. (We could explain random surface charges as due to anisotropy and variability of the (G-10) insulator but not surface charges that, in that part of the fine field cage, are always negative.)

Figure 16.9 shows the nature of the error potential, at least crudely, and how the electrons are believed to drift.

Recently, in a one-cubic-foot miniature TPC David Nygren, Rainer Sauerwine and Hiro Aihara have reproduced the "hook" phenomenon.

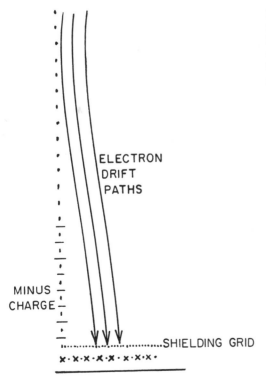

Figure 16.9. Electron drift trajectories, as calculated, showing a few sense wires near the left edge of a sector that receive no electrons. The sense wires are indicated by crosses.

ELECTRON
DRIFT
PATHS

MINUS
CHARGE

SHIELDING GRID

Their tests have shown that a fin-type fine field cage eliminates the effects of surface charges on the insulating surface. (See Fig. 16.10.)

Figure 16.11 shows the arrangement of wires of a sector of the TPC endcap. The electric field is generally upward, being stronger below the grid wires than above.

The sense wires (of 20 micrometers diameter) are operated at +3600 V. They are denoted by crosses in the figure. Gas multiplication at the sense wires is about 1000. The other wires are 75 micrometers in diameter. The field wires are operated at 800 V, to give the desired gain at the sense wires and at the same time promote mechanical stability of the sense wires under electrostatic forces.

A second cause of distortion was easily dealt with. In order to obtain a satisfactory uniform electric field the *average* potential in the grid-wire plane must match up with the fine-field-cage potentials to give a uniform electric field in the active volume. The average potential in the grid-wire plane can differ from the potential of the grid wires by 50 volts or more. This source of distortion was well observed in the one-cubic-foot miniature TPC, and we understand how to avoid trouble from this effect.

A third and serious source of distortion in track reconstruction, space-charge distortion, is due to positive ions. With a gas gain of 1000, there are potentially 1000 positive ions that can emerge from the end sectors for every one electron that drifts to the TPC endcaps. Since some positive ions are captured within the sector wires, the number of emitted positive ions may be somewhat less—say 500 for each electron. These positive ions move rather slowly (10 meters/sec) through the active volume, and, while in the active volume, they distort the electric field. Furthermore, the resulting distortion is time-dependent, so hard to correct for.

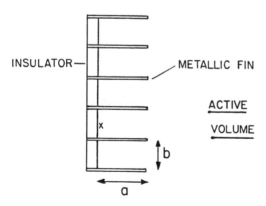

Figure 16.10. Fin-type field cage. Providing the dimension a is at least 1.5 times the dimension b, any erroneous potential on the insulator surface at x is suppressed by a factor of 50 to 100, according to electrostatic calculations.

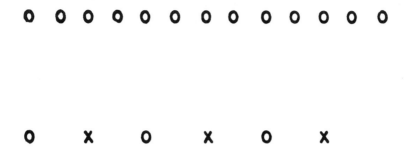

Figure 16.11. Arrangement of wires in a TPC end sector. The topmost row of wires is the (shielding) grid, with 1-mm wire spacing. In the next row of wires the sense wires are represented by crosses and the "field" wires are shown as circles. At the bottom of the figure there is a conducting ground plane. Proportional-counter action occurs at the sense wires, where the electric field is strongest. The active volume of the instrument is above the shielding grid.

The cure for space-charge distortions, which is to be implemented in the summer and fall of 1983, is to add a gating grid to the end-sector structure.

Figure 16.12 shows the planned structure as developed and tested by Peter Nemethy, Nobu Toge, and Peter Robrish. For normal operation all the gating-grid wires are set to the same potential that we will call V_0. When the gate is to be closed, alternate wires will be set to $V_0 +$ 150 V and $V_0 - 150$ V. Then electric field lines from the main active volume terminate on positive gating-grid wires, and electric field lines from the sense wires terminate on negative wires of the gating grid. Electrons from the main volume then cannot enter the region of gas multiplication, and positive ions cannot proceed from the sense wires into the main volume.

In operation, the gating grid would be closed until an interesting event was indicated by the triggering setup. Then, in 2 microseconds, the gate would be opened, staying open for about 20 microseconds.

The features that require careful development are:

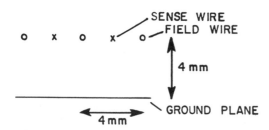

o o o o o o o o o o GATING GRID

Figure 16.12. End-sector wire arrangement redesigned to include a gating grid.

a. The gate must open fully within about 2 microseconds.
b. The amplifiers, one of which is connected to each sense wire and each pad, must be in condition to give usable information on pulse heights within about 2 microseconds. (In 2 microseconds the electrons drift about 10 cm; we do not want to lose more than 10 cm of the one-meter drift distance.)

We believe, judging by tests performed with a prototype pulser to form the gating potentials, that these criteria can be met, and we are very hopeful that they can be met within the physical limitations of the present TPC geometry. Then total electrons entering the gas-gain region around the sense wires could be reduced by perhaps a factor of 50 and the fraction of positive ions leaving the sense wires that enter the main volume could be reduced by a factor of about 50. It then seems possible to lower by a factor of 2500 the number of positive ions in the main volume.

It seems, then, that we have a suitable cure for all the observed distortions, but that much time and effort must go into the needed changes. In the meantime, we are looking at our results and learning how to compensate, at least in part, for the distortions that are (temporarily) present.

Results

The results now available are based on the running during May and June of 1982. Rapid progress is being made toward perfecting analysis programs that will keep the principal analysis effort up to date in "real time", but as of now (April 1983) the latest data recorded are not analyzed. For the May-June running the integrated luminosity is 4800 inverse nanobarns.

Figure 16.13 shows the PEP-4 (TPC) results for pion fraction among particles of hadronic events (as well as kaon and proton fractions) as a function of the particle momentum. For comparison, Figure 16.14 shows the more complete results from other experiments, mostly from TASSO at PETRA (in Hamburg). Clearly we have to perfect our (PEP-4) ability to distinguish among particle types.

Figure 16.15 shows the x distribution for charged pions, as seen in the PEP-4 results. x is the pion energy divided by its kinematic maximum, which is half the c.m. energy. Figure 16.16 contains other similar data,

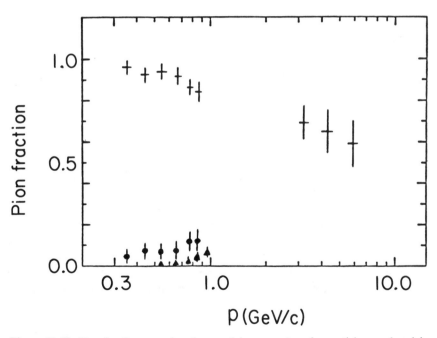

Figure 16.13. Pion fraction as a function particle momentum for particles produced in positron-electron collisions. Also shown are kaon fractions (circles) and protons (triangles). These are PEP-4 (TPC) results.

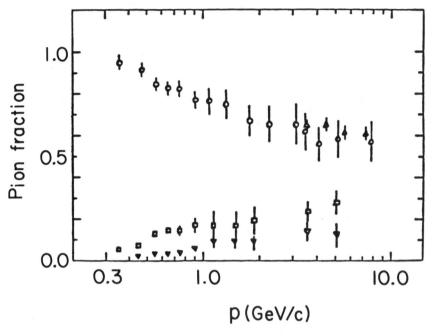

Figure 16.14. TASSO and DELCO results on pion fraction (and kaon and proton fractions) as a function of particle momentum. Kaon points are open squares. Proton points are inverted triangles.

mostly from TASSO at PETRA. These distributions bear on the question of how energetic quarks convert themselves into the physical particles that we actually observe.

The most common hadronic event in these data is a 2-jet event. Electron and positron annihilate into a single virtual photon which becomes a quark-antiquark (q-$\bar{q}$) pair. As the energetic quarks try to separate from each other a strong color field develops that finally "pulls out of the vacuum" a string (in momentum space) of particles that we see as particles in two jets, one jet in the direction of each original quark. It is characteristic that the particles in a jet have rather limited momentum components (say 300 MeV/c) perpendicular to the jet axis and a spread of momenta in the axis direction.

But some events appear to be 3-jet events, which are expected to be seen when one quark emits a gluon and it results in a third jet.

To analyze hadronic events it is customary to construct from the

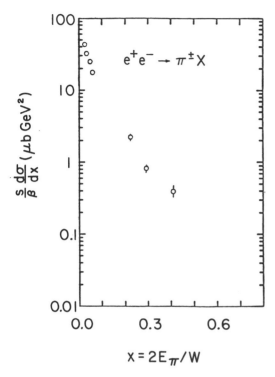

$$x = 2E_\pi / W$$

particle momenta a tensor:

$$M_{\alpha\beta} = \frac{\displaystyle\sum_i p^i_\alpha p^i_\beta}{\displaystyle\sum_i \sum_\gamma p^i_\gamma p^i_\gamma}$$

where p^i_α is the α component of the momentum vector of the ith emerging particle. This is so constructed that its trace is unity. When it is diagonalized by referring it to a rotated coordinate system it can take the form:

$$M_{\alpha\beta} = \begin{bmatrix} Q_3 & & \\ & Q_2 & \\ & & Q_1 \end{bmatrix} \quad \text{with} \quad Q_3 > Q_2 > Q_1 .$$

If the particle momenta are close to two oppositely directed jet axes this orthogonalized tensor has the form:

$$Q_3 \approx 1, \qquad Q_2 \cong Q_1 \cong 0 .$$

If there is a 3-jet event then in momentum space all the particles are

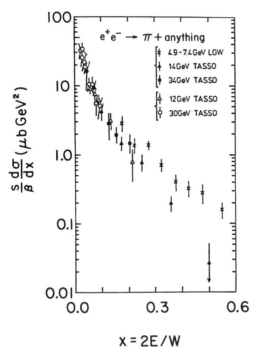

Figure 16.16. The quantity $(s/\beta)d\sigma/dx$ for pions, as reported by the TASSO collaboration at PETRA.

close to a plane (the plane that contains the momenta of the quark q, the antiquark $\bar{q}$, and the gluon g), and the tensor takes the form $Q_3 =$ large, $Q_2 =$ large, $Q_1 \cong$ zero. As a special case consider the momentum vectors coming out from the origin making a pancake shape; then $Q_3 = Q_2 = \frac{1}{2}$, $Q_1 = 0$.

If the momenta fill all directions rather uniformly then the values will be $Q_3 = Q_2 = Q_1 = \frac{1}{3}$. In any case we are assured by the definition of our tensor that $Q_1 + Q_2 + Q_3 = 1$.

The customary expressions of these properties involve the quantity called sphericity S, defined by

$$S = \tfrac{3}{2}(Q_1 + Q_2), \quad \text{such that } 0 \leqslant S \leqslant 1;$$

and the quantity called aplanarity, defined as

$$\text{aplanarity} = (\tfrac{3}{2})Q_1, \quad \text{implying } 0 \leqslant \text{aplanarity} \leqslant \tfrac{1}{2}.$$

These quantities can be portrayed as points in a two-dimensional diagram, since Q_3 is known once Q_1 and Q_2 are given, as $Q_3 = 1 - Q_1 - Q_2$.

Figure 16.17 shows the usual representation. Each hadronic event is a dot in the diagram. The concentration of dots in the left corner represents

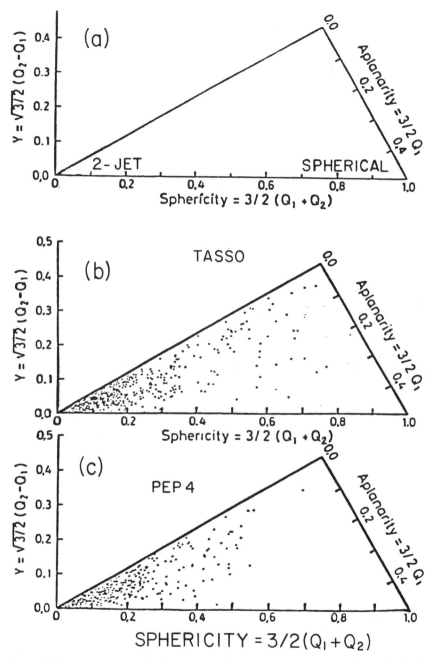

Figure 16.17. Representation of sphericity and aplanarity, each dot standing for one hadronic event and each dot falling inside the given triangle. (a) The diagram labeled to show 2-jet events in the left corner, typical 3-jet events along the upper-left side of the triangle and very spherical events in the right (lower) corner region. (b) TASSO and (c) PEP-4 (TPC) results.

the 2-jet events. The dots along the upper edge show clear 3-jet events. The presence of dots in other parts of the triangle could indicate either 4-jet events or a structure other than jet structure. Obviously we will be looking into the theory of these jets, hoping to determine whether existing theory gives an explanation for the observed density of dots in various parts of the triangle.

A number of other results have come out of our running, including values for R, the ratio of hadronic final states to μ^+-μ^- final states, and α_s, the strong coupling constant. Again, our results are consistent with the results of others; as yet our accuracy is less.

To summarize, we are at the point where we are starting to get real physics results, but in our first real running we are mostly observing things that are already known. We expect to get into new territory as we sharpen up our ability to distinguish particle types by utilizing fully the special ability of the TPC to measure specific ionization. Spatial distortions all have known cures which we expect to implement during 1983. By 1984 we expect to have a powerful new instrument in operation.

Getting this new-style detector in smooth operation seems to take much tender loving care, but we expect the final product to be well worth the time and effort spent.

17 Remarks on the Future of Particle Physics

Robert Serber

The development of particle physics has been closely tied to the construction of larger and larger particle accelerators. We are now reaching trillion-electron-volt rings, and one might consider as possible another increase of a factor ten. But it is difficult to believe, beyond that, in yet another ten, unless some new method of accelerating particles is invented. Experimental particle physics need not rely on ever larger accelerators. There is another road: learning by studying the fine effects, by ever more precise and sensitive measurements. That method, though, requires a theory capable of interpreting the results.

Another field for the future involves relativistic heavy ions. There have been suggestions that, in the collision of relativistic heavy nuclei, novel cooperative phenomena might appear.

A number of predicted particles have not yet been found. The oldest dates back to 1916. While we are far from being able to do Compton scattering with gravitons, the detection of gravitational waves may not be so far off. Present experimental sensitivity is within a factor of one hundred of that thought needed to detect the emissions from the nearer galaxies.

At the beginning of the century we had an elaborate spectroscopy and no theory which could interpret it. Today we also have an extensive

NEW DIRECTIONS IN PHYSICS
The Los Alamos 40th Anniversary Volume
ISBN 0-12-492155-8

spectroscopy, the masses of all the observed particles, and much other data. We also have a plausible theoretical structure. The difficulty lies in being able to calculate accurately the consequences of the theory. There will undoubtedly be great efforts in many directions to improve this situation. One of the promising methods is to use computers to make quantum chromodynamics calculations on a lattice. People are already beginning to build computers specifically designed to solve QCD lattice problems. It seems possible that such computers could be a million times faster than the general purpose computers used now.

Nature has given us no hint of a breakdown of quantum mechanics, even at distances of 10^{-16} cm. Our only outstanding embarrassment is the failure to reconcile quantum mechanics and the general theory of relativity. Even without new physical clues, mathematical studies of possible generalizations of quantum field theory will be pursued. For example, T.D. Lee and his collaborators have been considering relativistic field theories which incorporate a fundamental length. If such theories gave better convergence it might be possible to extend the set of theories which can be considered to include non-renormalizable theories.

Finally, let me quote from Einstein: "In a sensible theory there are no dimensionless numbers whose values are determined only empirically. I can, of course, not prove that dimensionless constants in the laws of nature, which from a purely logical point of view can just as well have other values, should not exist. To me in my faith in God this seems evident, but there might be few who hold the same opinion." Our present theoretical structure contains a couple of dozen such dimensionless constants. With sufficient faith in God, an optimist might hope sometime to return to the situation of the good old days when the only thing we didn't understand was the fine structure constant.

18 Supernova Theory

H.A. Bethe

Introduction

Supernovae are observed by their enormous luminosity which starts suddenly and continues for about one year. The luminosity is of the same order as that of an entire galaxy. In addition to this copious emission of light, material is expelled from a supernova that is afterwards observable as a nebula like the Crab Nebula. The kinetic energy in this nebula is greater than the total light emitted. The order of magnitude of the energies involved is 10^{51} erg for which we introduce the abbreviation *foe* (fifty-one erg). It is generally accepted that supernovae of Type II arise at the end of the evolution of very massive stars, of masses about 10–30 $M_\odot$. In the simplest theory, which we shall discuss through most of this paper, the center of this star collapses to a density about equal to nuclear density. A rebound then occurs that drives a shock out through the surrounding material, and this shock leads to the expulsion of most of the mass of the star.

The majority of the work reported here was done by Gerald Brown and me during the period 1978–83. We had several collaborators, first J. Applegate and J. Lattimer, later Gordon Baym, and recently J. Cooperstein. We relied heavily on computer calculations by James Wilson of

NEW DIRECTIONS IN PHYSICS
The Los Alamos 40th Anniversary Volume
ISBN 0-12-492155-8

Livermore, by David Arnett of Chicago, and by J. Cooperstein. Many ideas used here are due to Amos Yahil. The initial, pre-supernova configuration was taken either from the work of Weaver, Woosley, and collaborators [1,2], or from Arnett [3].

Pre-supernova Configuration

Figure 18.1, due to Arnett, shows the distribution of density in a highly evolved star just before it becomes a supernova. The shells are marked in which nuclear reactions take place, involving He, C, Ne, O and Si. The temperatures at which these reactions take place are given in units of 10^9 degrees.

Inside the shell in which Si reacts, there is a sphere in which the Si reaction has gone to completion, and which therefore consists of Fe and related elements. This sphere is supported by degenerate electron pressure. Since densities are high, the electrons are relativistic, particularly near the center. The mass of this central sphere is therefore limited by the

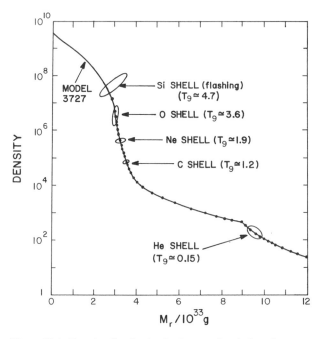

Figure 18.1. Density distribution in the state just before the supernova implosion. Abscissa is the included mass, ordinate the density in g/cm³. Temperatures in billions of degrees are noted.

Chandrasekhar mass which is

$$M_{Ch} = 5.76 \, Y_e^2 M_\odot \left[1 + \frac{3}{2} \left(\frac{\pi T}{\mu_e} \right)^2 - \dots \right] \tag{1}$$

where $M_\odot$ is the mass of the sun, Y_e is the electron fraction, i.e., the number of electrons per nucleon, T stands for kT with k the Boltzmann constant, and μ_e is the Fermi energy of the electrons. The expression outside the bracket is the well-known Chandrasekhar mass for a cold electron gas, the bracket gives the enhancement due to electron temperature. This can conveniently be expressed in terms of the entropy. We generally use S to denote the total entropy per nucleon in units of the Boltzmann constant, and by S_e the fraction of the entropy which is in electrons. In terms of S_e, Eq. (1) may be rewritten

$$M_{Ch} = 5.76 \, M_\odot \left[Y_e^2 + \frac{3}{2} \left(\frac{S_e}{\pi} \right)^2 - \dots \right]. \tag{2}$$

The mass of the Fe sphere increases gradually to M_{Ch} as the Si reacts; this process takes a few days. Once M_{Ch} is reached, the core of the star becomes unstable and contracts adiabatically. In this process, the temperature increases, especially at the center, and α-particles and neutrons are released from the nuclei. This release takes energy, and the energy is taken away primarily from the biggest reservoir of energy, the thermal energy of the electrons. The electron entropy therefore decreases; we may say that paradoxically the electron gas cools down in the process of the general adiabatic increase of temperature. Using (2), the Chandrasekhar mass decreases. Initially, the bracket in (1) is about 1.2, but when the density reaches 10^{10}, it has decreased to about 1.02. This means that the material has become very unstable, and the core collapses rapidly. Using computer calculations, this collapse takes a fraction of a second.

Adiabatic Collapse

The collapse of the core is essentially adiabatic, i.e., the entropy of any material element remains approximately constant. Some deviation from this is due to capture of electrons which will be discussed later, but this does not change the characteristics of the collapse in a major way.

It has been shown by Goldreich and Weber [4], and by Yahil and Lattimer [5] that the inner part of the core collapses homologously. This means that the distribution of density remains essentially unchanged, but all densities are increasing by the same factor which depends only on time. The mass included in this "inner core" is approximately the Chan-

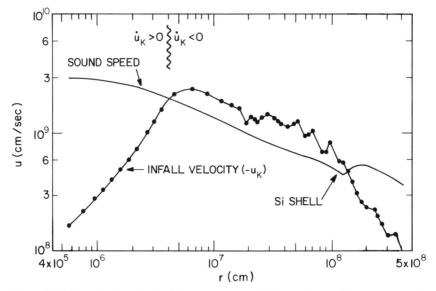

Figure 18.2. Distribution of velocities at some stage in the collapse of a supernova. The material inside of 40 km is the homologous core, outside of this is the material collapsing with super-sound velocity.

drasekhar mass of a cold electron gas (because now S_e is small) of the proper electron fraction Y_e. Details are given especially in [5].

Outside the inner core, the material moves supersonically, i.e., the infall velocity u is greater than the local sound velocity. In this outer core it can be shown that the infall velocity is approximately a constant fraction of the free fall velocity

$$u = \alpha \left[\frac{2GM(r)}{r} \right]^{1/2} \tag{3}$$

where α is usually of the order of $\frac{1}{2}$, and depends on the value of Y_e. $M(r)$ is the mass enclosed by radius r. The density of the material in the region is expressed roughly by the relation

$$\rho = C \cdot 10^{31} \cdot r^{-3}. \tag{4}$$

The occurrence of the power 3 in this equation is related to the equation of state of a relativistic electron gas, which is

$$p \sim \rho^{4/3}. \tag{5}$$

The constant C in (4), according to most computer calculations, is about three.

A "snapshot" of the distribution of velocity u vs. r is shown in Figure 18.2, due to Arnett [3].

Electron Capture

The electrons have high Fermi energy, roughly

$$\mu_e = 11 \, \text{MeV} \, (\rho_{10} \, Y_e)^{1/3} \tag{6}$$

where ρ_{10} denotes the density of the material in units of 10^{10} g/cm^3. Because of this high electron energy, it is energetically favorable for the electrons to be captured by nuclei, increasing the number of neutrons in the nuclei at the expense of the number of protons (neutronization). The capture of electrons can occur either on free protons or on protons bound in nuclei.

The capture on free protons is

$$p + e^- = n + \nu. \tag{7}$$

Its rate (per electron) is proportional to the number of protons per unit volume. This number is very small and is related to the number of neutrons. We denote the number of free protons per total nucleon by X_p, and the number of free neutrons by X_n; then

$$X_p = X_n e^{-\hat{\mu}/T}. \tag{8}$$

Here

$$\hat{\mu} = \mu_n - \mu_p \tag{9}$$

denotes the difference between the chemical potentials of neutron and proton. Since the nuclei are already rich in neutrons, $\hat{\mu}$ is positive and can become quite large; a rough approximation is

$$\hat{\mu} = 200(0.45 - Y_e). \tag{10}$$

The temperature T is generally of the order of 1 MeV so that X_p/X_n is quite small. The neutron fraction X_n is itself quite small, usually of the order of a few percent, and the proton fraction is very small, commonly 10^{-5}–10^{-6}. Therefore capture by free protons is a relatively rare process.

For this reason, BBAL [6] concluded that electron capture takes place mostly on complex nuclei. They estimated the rate of these captures by assuming that the electron capture is an allowed weak interaction transition. Assuming this, they got a fairly rapid capture of electrons and found that at density 5×10^{11} the electron fraction is about

$$Y_L = 0.32, \tag{11}$$

a rather small number. Accordingly, the Chandrasekhar mass in this gas would be quite small. It will be shown later on that this would make it very unlikely that the outgoing shock will succeed in getting to the outer envelope of the star.

Fortunately, Fuller [7] pointed out an error in the work of BBAL. Figure 18.3 shows a schematic of the energy levels of neutrons and protons in the neighborhood of Fe^{56}. BBAL assumed that a proton from the $f_{\frac{7}{2}}$ shell can go into the nearly empty $f_{\frac{5}{2}}$ shell of neutrons. This is an allowed Gamow-Teller transition. Fuller pointed out that the $f_{\frac{5}{2}}$ shell is filled when there are 38 neutrons in the nucleus. This occurs rather early, at a density of approximately 10^{11}. At higher densities, this allowed transition is no longer available, and the electron capture must instead take place by a forbidden transition, leading for example to the $g_{\frac{9}{2}}$ level of neutrons. The probability of this is very much lower than that of an allowed transition. Accordingly, Fuller, Cooperstein [8] and others concluded that the electron fraction at a density of 5×10^{11} is approximately

$$Y_L = 0.39. \tag{11a}$$

This gives a much larger Chandrasekhar mass, which is favorable for a successful shock.

The density 5×10^{11} has been chosen because at this density neutrinos begin to get trapped. This is due to the weak neutral currents which cause scattering of neutrinos by nuclei, especially those rich in neutrons; see [6].

After neutrinos are trapped, electron capture still takes place, but the neutrinos resulting from that capture will stay inside the core of the star. The Y_L in (11a) then denotes the total lepton fraction, i.e.,

$$Y_L = Y_e + Y_\nu. \tag{12}$$

When the density reaches about 10^{13}, electrons and neutrinos are in equilibrium so that the following relation holds between their chemical potentials:

$$\mu_e - \mu_\nu = \hat{\mu}. \tag{13}$$

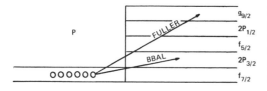

Figure 18.3. Schematic levels of protons and neutrons in a typical nucleus. The lower arrow denotes the electron capture transition postulated by BBAL [6], the upper arrow that postulated by Fuller [7].

If Y_L = .4, then ultimately, at a density of 10^{14} or higher, we have approximately

$$Y_e = .32, \qquad Y_\nu = .08. \qquad (13a)$$

Equation of State

It is essential for these calculations to have a good equation of state for the nuclear component of the matter. Such an equation was derived by Lamb, et al. [9], and is shown in Figure 18.4. The dotted lines are adiabats for various entropies S. The big dashed line is the condition in which half the material is in complex nuclei and the other half in free nucleons. It is seen that the adiabat $S = 1$ is consistently far below the dashed line so that nearly all the nuclear material is in complex nuclei. The entropy in this case is predominantly in excitation of these nuclei. Very little pressure is caused by the nuclei; essentially all the pressure is due to leptons.

Slightly above a density of 10^{14}, there is a shaded region in which the material consists of nuclear matter at slightly reduced density with bubbles

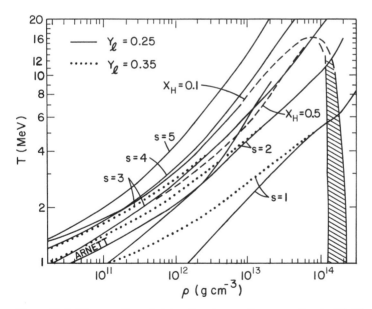

Figure 18.4. The equation of state of nuclear matter, according to [5]. The dotted curve with $S = 1$ is close to that relevant for our collapse. The curve $X_H = 0.5$ is the line at which half the nuclear material is in complex nuclei, the other half in nucleons.

of vacuum enclosed. While this is interesting scientifically, it has little influence on the relation between pressure and density along an adiabat. When the adiabat continues beyond normal nuclear density, it represents compressed nuclear matter. In this region, the pressure increases very rapidly with density, roughly

$$p = p_0 \left(\frac{\rho}{\rho_0} \right)^3. \tag{14}$$

Thus, for $\rho > \rho_0$, the material becomes highly incompressible.

The equation of LLPR [9] was based on the Skyrme interaction which is apt to be somewhat "too hard", i.e., to give too high pressure for a given density when $\rho > \rho_0$. BBAL modified the LLPR equation of state accordingly.

A slightly different equation of state was derived by BBCW [10]. At densities well below nuclear, the BBCW equation is closer to simple physical considerations than LLPR. For $\rho > \rho_0$, it is essentially the same as BBAL. The transition between these two regions was accomplished by Cooperstein [8] and could probably still be made somewhat smoother.

Collapse and Start of Shock

As was discussed earlier, the inner core collapses homologously as long as the density is not too high. When the density at the center reaches nuclear density, the material becomes nearly incompressible. Therefore, the infall is stopped, first at the very center and then rapidly at points farther out. Let R_1 be the outermost point at which nuclear density is reached. From this point pressure waves will be emitted which go through all the material farther out. We shall call the material at nuclear density and higher the "dense pack".

The pressure wave that signals the existence of the dense pack brings the material farther out essentially to rest. It amounts to a sudden change of velocity from u to approximately 0. Pressure waves with a sudden change of velocity are shock waves, so we have here the start of a shock. In any pressure wave, including shock waves, there is a change of density related to a change of velocity; this relation is

$$\frac{\Delta \rho}{\rho} \simeq \frac{\Delta u}{a}, \tag{15}$$

where a is the local sound velocity.

With this change of density and velocity, there is also connected an increase in entropy, which is

$$\Delta S \simeq 0.3 \left(\frac{\Delta u}{a} \right)^3. \tag{16}$$

This entropy change is very small as long as u is small compared to a, but becomes sizable when $u \simeq a$. We can therefore say that the shock effectively starts at that point which is the sonic point of Figure 18.2.

Beyond the sonic point, the entropy behind the shock increases. Figure 18.5 shows this behavior from a schematic calculation by Yahil [11]. In this calculation, Yahil used a new numerical method, based on a method of the Naval Research Laboratory, in which shocks appear as sharp discontinuities; i.e., the method avoids artificial viscosity. It is seen in Figure 18.5 that the entropy rises first very slowly with the enclosed mass M, and then very rapidly, to reach about $S = 6$ just behind the shock. This is typical of other calculations. Of course, every mass element

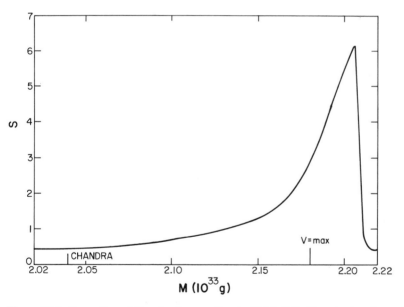

Figure 18.5. Formation of the shock according to Yahil [11]. The abscissa is enclosed mass, the ordinate is entropy. The unperturbed mass had an entropy of about 0.4. One line marks the mass of the Chandrasekhar core for the appropriate Y_e, another line is the position of maximum infall velocity. Note that the scale covers only about 10% variation of enclosed mass.

that has been traversed by the shock thereafter keeps the entropy it has acquired in the shock, unless it is subsequently hit by a second shock.

The formation of a shock was first shown in computations by Wilson [15]. As has been explained, the shock forms approximately at the sonic point, actually somewhat further out [5]. Thus it encloses an inner core of mass approximately equal to the Chandrasekhar mass for the appropriate Y_e. The higher Y_e, the larger the mass of the inner core. The inner core remains essentially unshocked and retains the entropy it had to begin with.

Energetics

A star originally has a small energy. In particular, the Fe core has usually an energy of about $-.1$ to $-.2$ foe (1 foe $= 10^{51}$ erg). In the collapse, large amounts of gravitational energy are set free, and most of this is converted into internal energy. It is useful to consider the total energy per unit mass (specific energy)

$$\varepsilon = \frac{-GM}{r_M} + \varepsilon_{int} + \frac{1}{2}u^2. \tag{17}$$

The first term here is the gravitational energy, the second term the internal energy, and the last the kinetic energy. This last term is usually not very important. The internal energy in turn consists of the energy in leptons, the energy of a cold nuclear gas of the given density, and the thermal energy. We shall be especially interested in the total energy of a given mass of material

$$E = \int_a^b \varepsilon(M)\, dM. \tag{18}$$

In order for the shock to be successful, the total energy in the shocked region should be positive. Since the total energy of the entire star is initially near zero, we should therefore require that the energy in the inner (unshocked) core is negative. In other words, E as defined in (18) should be negative if $a = 0$ and b is the radius of the unshocked core, R_2.

This does not yet guarantee that the energy in the shocked part is positive because energy is lost in the form of neutrinos. The total neutrino loss in a reasonably successful shock is about

$$E_\nu \simeq 2 \text{ foe}. \tag{19}$$

Of this, about 0.3–0.8 foe is lost during the collapse and the rest during the expansion. Some details of this will be discussed later.

When talking of the total energy of the star, it is convenient to take the outer limit to be the surface of the Fe core. The pressure at this surface, R_3, is quite small, and also it moves with small velocity. Therefore rather little energy is transmitted through R_3 from the mantle of the star to the core. Below (18), we considered the energy of the unshocked core. It is therefore reasonable to take the shock energy to be the region from R_2 to R_3. This then includes the shocked region and also the part of the core that is still falling in, i.e., the part between the shock and R_3. The latter region has a negative energy because in the gravitational infall work is done by the outside on the inside. This material of negative energy, however, is gradually swallowed up by the shock, without any change of the energy of the material. Therefore a realistic value for the energy of the shock can only be obtained if the outer, still infalling material is included.

Figure 18.6 gives the distribution of the specific energy in a schematic calculation by Yahil [11]. The innermost part of the core is omitted; it has a positive specific energy because the gravitational energy goes to

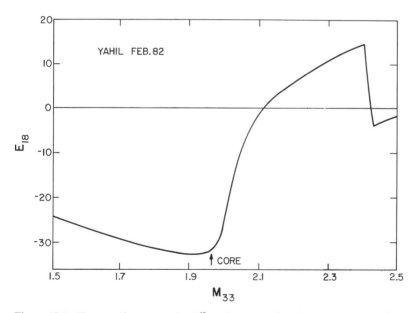

Figure 18.6. The specific energy, in 10^{18} erg/gm, as a function of the enclosed mass (in units of 10^{33} gm), according to Yahil [11]. To the left of the arrow is the unshocked core, to its right the shocked material. At the far right is the material not yet reached by the shock.

zero at the center of the star while the internal energy is positive. The outer part of the unshocked core, however, has a negative specific energy because here the negative gravitational energy gives the greatest contribution. The innermost part of the shock still has a negative energy, as the figure shows. However, as the entropy in the shocked material rises, so does its internal energy, and so from a certain point, R_4, on out the specific energy is positive. The point R_4 does not move much in mass as time goes on. It is therefore convenient to define the shock energy as the energy contained between R_4 and R_3.

The specific energy rises to a maximum just behind the shock and then falls discontinuously to a negative value in the infalling material, as has been previously discussed.

Estimate of the Energy of the Inner Core

Cooperstein [12] has pointed out that the energy of the inner core can be estimated fairly well by assuming it to be composed of two polytropes. The inner polytrope, up to R_1, has a high adiabatic index γ_N where N stands for nuclear. This region is assumed to have a density equal to or higher than nuclear density ρ_0. Using

$$p \sim \rho^{\gamma_N}, \tag{20a}$$

we have $\gamma_N = 3$.

Between R_1 and R_2, the pressure is primarily due to electrons, and therefore we have a gas of adiabatic index $\gamma = \frac{4}{3}$. Now assuming the entire inner core to have come to rest, then in each of the two regions the Virial Theorem [13] will hold:

$$V_N + 3W_N = 4\pi p_1 R_1^3$$
$$V_e + 3W_e = 4\pi p_2 R_2^3 - 4\pi p_1 R_1^3$$
$$W = \int \left(\frac{p}{\rho}\right) dM \tag{20}$$
$$V = -G \int \left(\frac{M}{r}\right) dM$$

Using the relation between pressure and internal energy, the Virial Theorem (20), some property of polytropes, and a good deal of algebra, one can deduce the following result:

$$E - 4\pi p_2 R_2^3 = V_N \frac{3\gamma_N - 4}{3\gamma_N} + E_X \tag{21}$$

where E_X is the energy of internal nuclear excitation. The gravitational potential energy in the dense region denoted by N is very accurately the same as that of a sphere of uniform density, i.e.,

$$V_N = -\frac{3}{5} G \frac{M_N^2}{R_1}.$$ (22)

Deducing M_N and R_1 from the computer output of Cooperstein, we find from (21) and (22)

$$E - 4\pi p_2 R_2^3 = -8.0 \text{ foe} + E_X.$$ (23)

More detailed use of the computer output gives for the first term of the right-hand side

$$-7.5 \text{ foe.}$$ (23a)

It is interesting that the theory does not give the energy of the inner core directly, but rather the difference between E and the quantity

$$4\pi p_2 R_2^3.$$ (23b)

We shall call

$$4\pi p_2 R_2^3 - E$$ (24)

the net ram pressure (NRP). As (23) shows, the NRP has a large positive term, which arises from the dense center, and a negative term due to the excitation of the nuclei.

The ram pressure is essential to drive the shock. Therefore, the greater the NRP the better is the chance that the shock will propagate. A high energy of the dense center is clearly very helpful, while excitation of nuclei is harmful. Other harmful terms will be discussed later, and we shall find later that the ram pressure remaining at long distance is very small.

It is therefore essential that E_X be kept as small as possible. One can show that the excitation energy per unit mass is

$$\varepsilon_X \simeq \tfrac{1}{2} S_X T$$ (25)

where S_X is the part of the entropy which is in nuclear excitation, and T is the temperature. Most of the entropy is actually in nuclear excitation. Evaluation shows that

$$\varepsilon_X \simeq 1.6 \cdot 10^{18} \cdot S_X^2 \text{ erg/g.}$$ (26)

The inner core contains about 1.6×10^{33} g, so that the total energy in excitation is

$$E_X = 2.6 S_X^2 \text{ foe.}$$ (27)

It is desirable to keep $S_X < .6$; then the excitation energy is less than 1 foe.

Two Calculations by Cooperstein

On the basis of the arguments in the last section, Cooperstein has run two computer calculations with very low starting entropy. One calculation, in which the initial entropy near the center was $S_i = 0.5$, gave a successful shock. The other, with an initial entropy $S_i = 0.7$, gave a shock which started out reasonably but which then petered out at a radius R of about 120 km. Figure 18.7 shows the distribution of specific energy in Cooperstein's successful shock as a function of time. The time, in arbitrary units, is noted on the curves. As in Yahil's calculation (Fig. 18.6), ε is negative in the innermost part of the shocked region, rises quickly, and drops abruptly at the shock itself. The shock spreads over a wider region of mass as time goes on, as expected. The specific energy just behind the shock decreases accordingly because the total energy in the shock remains roughly constant. At the last recorded time, 9293, the shock has

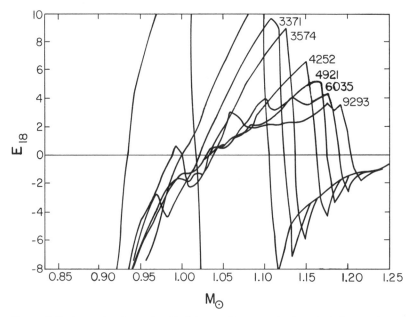

Figure 18.7. Successive "snapshots" of the specific energy in the calculations by Cooperstein [12]. The abscissa is enclosed mass in units of the solar mass, ordinate the specific energy in units of 10^{18} erg/cm. The curves are labelled by the time in arbitrary units.

progressed to a mass $M \simeq 1.2 \, M_\odot$ which is close to the outer edge of the Fe core at $1.25 \, M_\odot$. There is every reason to expect that the shock will propagate successfully to the edge of the Fe core and beyond. If it gets as far as the shell of O^{16} which is not far outside the Fe core, further energy will be added to the shock by nuclear synthesis in which a large part of the O^{16} is converted into Fe.

Figure 18.8 shows the development of the Cooperstein unsuccessful shock. It starts out much the same as the successful one, but very quickly the specific energy behind the shock front decreases greatly, and by time 3858 it has become negative. It is almost certain that this shock will be a failure.

It is rather remarkable that such a small difference in initial entropy can make the difference between success and failure. Some insight into the sensitivity is given by a more detailed consideration of the Virial Theorem which we rewrite as follows:

$$4\pi p R^3 - E = \int \left[3\left(\frac{p}{\rho}\right) - \varepsilon_i \right] dM - \int r\ddot{r} \, dM. \qquad (28)$$

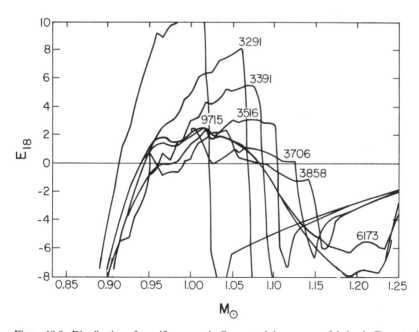

Figure 18.8. Distribution of specific energy in Cooperstein's unsuccessful shock. For notation, see Figure 18.7.

This equation is valid between any limits. The last term, involving the acceleration, is quite difficult to discuss, and we shall therefore consider only the part of the star which has become hydrostatic at some given time. We shall do this for a rather late time when the shock has progressed to a distance $R_S = 400$ km, and we shall use Cooperstein's successful shock computation.

We shall distinguish five regions:

1. The dense core, $\rho > \rho_0$. This gives a large positive integral, about 7.5 foe as we previously mentioned.

2. The inner core, $\rho < \rho_0$, $S < 2.7$. In this region we have complex nuclei. The integral on the right gives a negative contribution for two reasons, (a) excitation of nuclear levels and (b) negative nuclear pressure that is due to the decreasing Coulomb energy, see [10]. The limit of $S = 2.7$ is chosen such that half of the nuclear matter is in complex nuclei and the other half in nucleons.

3. The inner shock, $2.7 < S \leqslant 8$. Here we have free nucleons. The integral on the right gives a very small contribution. There is a positive component because for free nucleons the thermal energy of the nucleons is given by

$$\varepsilon_{\text{th}} = \frac{3}{2} \cdot \frac{p}{\rho}. \tag{29}$$

But this relatively small energy is balanced by the energy required to dissociate nuclei into nucleons, $Q_D \simeq 8.8$ MeV for Fe. The integrand is exactly zero when

$$T = \frac{2}{3} \cdot 8.8 \text{ MeV} \simeq 5.9 \text{ MeV}, \tag{30}$$

and this is approximately the temperature in this inner shock.

4. Main shock. The entropy S is about 8, the acceleration is taken to be 0. Here the first integral in (28) is strongly negative because

$$T < 5.9 \text{ MeV}. \tag{30a}$$

The loss due to dissociation is greater than the gain due to the p/ρ of free nucleons.

5. The shock front region. Here $\ddot{r} < 0$. The second integral on the right of (28) is then positive, and therefore the NRP increases again. However, we have no good way to calculate this last integral; therefore, we shall leave out Part 5.

Table 18.1 gives the contribution of various terms in various regions to the NRP. The result is about 2.0 foe. The E of the entire Fe core at

Table 18.1 Energetics of Cooperstein's successful shock contributions to net ram pressure, $4\pi pR^3 - E$, in foe.

1.	Dense core		$+7.5$	
2.	a.	Excitation of nuclei	-2.5	
	b.	Negative nuclear pressure	-1.3	
		Total for Inner Core		3.7
3.	Inner shock region		0	
4.	Dissociation in main shock		-1.7	-1.7
		Total Net Ram Pressure		2.0
		Total energy of Fe core (neutrino losses)		-1.8
		Ram pressure, $4\pi pR^3$ (behind shock)		0.2

this time is about -1.8 foe because this much energy has been lost in the form of neutrinos. This gives a calculated ram pressure

$$4\pi pR^3 \simeq 0.2 \text{ foe}, \qquad (31)$$

which agrees reasonably well with the result of Cooperstein's computation. Actually, the computation shows about 0.4 foe; this difference is due to the fact that the region enclosed by the shock still has a somewhat higher energy than -1.8 foe because the region ahead of the shock has a negative energy.

The question then is how the shock can proceed with such a small ram pressure. We have the Hugoniot equation,

$$(u_2 - u_1)^2 = \frac{p_2}{\rho_1}, \qquad (32)$$

where subscript 2 refers to conditions behind, subscript 1 to ahead, of the shock. We can rewrite this in the form

$$(u_2 - u_1)^2 = \frac{p_2 R^3}{\rho_1 R^3}. \qquad (32a)$$

The denominator, relating to the density outside the shock, is given by

$$\rho_1 R^3 = C \cdot 10^{31}. \qquad (33)$$

Using $4\pi p_2 R^3 = 2 \cdot 10^{50}$ (Table I), if we had $C = 3$, as is the case in most of the infall, we would get

$$(u_2 - u_1)^2 \simeq 5 \times 10^{17}. \qquad (33a)$$

This would give a negative u_2 because the infall velocity $u_1 \simeq -10^9$. Fortunately, however, by this time C has decreased to about 0.8. This permits the shock to proceed.

Thus, it appears that the low density at large R is important for the success of the shock, in addition to the low initial entropy. The low density, $C \simeq 0.8$, is partly already present in the initial conditions, cf. [12].

The loss of energy by dissociation is limited by the fact that below about $T = 1.5$ MeV, full dissociation stops. Using the Saha equation one finds that dissociation then goes only down to alpha particles. The dissociation energy is reduced to about 2 MeV per nucleon.

New Pre-supernova Calculations

The low entropy in Cooperstein's computations was an *ad hoc* assumption, in order to get a successful shock. Fortunately, just at the right time Weaver, Woosley, and Fuller [2] revised their calculations of the pre-supernova conditions. They took better account of the nuclear reactions which occur in the region between Mg and Fe, and found that the new chain develops less entropy. They also considered carefully the energy losses due to neutrino emission during the stages immediately preceding the supernova. As a result, they find for the center of star

$$S_i = .7 \tag{34}$$

whereas their old calculation gave

$$S_i = 1.0.$$

Moreover, their new assumptions lead to a low initial Y_e; approximately $Y_{ei} = .42$ at the center whereas the older calculations gave about .44. This has two desirable consequences: (1) There will be less electron capture from the start of the supernova collapse to the trapping of neutrinos, which means less loss of energy in neutrinos in that period, and, (2) the mass of the Fe core is much smaller,

$$M_{core} \simeq 1.35 \, M_{\odot}, \tag{35}$$

while the previous value was about 1.5. This means that the shock wave can reach the surface of the Fe core more easily, and once it proceeds into the Si shell it finds much lower density and can progress still more easily.

The new value $S_i = .7$ still makes the Cooperstein calculation fail. However, it is a big step in the right direction, and further changes may not be excluded. The final pre-supernova configuration may lead to a successful shock.

The Mine Field

Neutrinos remain trapped in the interior until the density outside the shock has decreased to about 10^{11}. After this, the previously trapped neutrinos in the inner part of the shock are released. The sphere at which this release takes place is called the neutrinosphere, and is at about 60–80 km. The release is by essentially black body radiation and takes a few milliseconds.

The neutrinos trapped in the inner core take much longer to escape. The best estimate is in a paper by Burrows et al. [14]. It takes about one second for the inner core to be deleptonized. During this time, the order of 50–100 foe is emitted in the form of neutrino energy.

The region which has densities between 10^{10} and 10^{11} when the shock reaches it is subject to considerable energy loss by neutron capture. As the shock arrives, the nuclei are dissociated into nucleons. Electrons are eagerly captured by the protons, giving neutrinos. Y_e goes down from something like .4 to something like .2 in this process. Most of the electron energy goes to neutrinos, which can escape easily. It can be estimated that this means an energy loss of about 1.5×10^{18} erg/g, and that the region affected is about .1 $M_\odot$ provided the shock is successful and proceeds reasonably rapidly through this density region. This amounts to an energy loss of about .3 foe in neutrinos.

However, if the shock already has become very slow, as a result of dissociation, then this energy loss may stall it completely. In this case, further infalling material will simply accrete to the existing shocked region. Thus, the density region around 10^{10} constitutes a "mine field", with considerable danger that the shock be converted to an accretion shock. This is probably the explanation why Cooperstein's computation with an initial entropy of .7 led to failure. In any case, it helps to explain the extreme sensitivity of the shock to initial conditions.

Wilson's Delayed Shock

James Wilson of Livermore has done two computations [15] in which the shock seemed to stall for a considerable time in the mine field. But then, after a time of .3–.5 seconds, the shock got going again.

The first such calculation was done for a star of mass 10 $M_\odot$. In the pre-supernova condition, the density in this star falls much more rapidly than the usual R^{-3}. Near the surface of the iron core, the density behaves about as R^{-5}. Consequently, the ram pressure $p_2 R_2^3$ needs to be only quite small, and can then still drive the shock according to the Hugoniot relation (32).

It is likely that Wilson's successful shock is mostly due to neutrinos coming out of the inner core. As has previously been stated, there are about 50–100 foe of such neutrinos per second. Perhaps 10 foe per second is in electron neutrinos and a similar amount in antineutrinos. These can be captured by nucleons as follows:

$$\nu_e + n \rightarrow e^- + p, \qquad \bar{\nu}_e + p \rightarrow e^+ + n, \qquad e^+ + e^- \rightarrow 2\gamma. \quad (36)$$

The neutrinos, coming from the inside, have a higher temperature (typically about 4 MeV) than the electrons (typically about 2 MeV); therefore the neutrinos transfer energy to the matter. This transfer, per nucleon, is independent of the matter density. Using the cross section of the reactions (36), the energy transfer from neutrinos to matter per nucleon is

$$\frac{d\varepsilon}{dt} = \frac{J_{51}}{r_7^2} \frac{\text{MeV}}{\text{nucl-sec}} \quad (37)$$

where r_7 is the distance from the center in units of 10^7 cm, and J_{51} is the energy current in neutrinos in foe per second. The energy in $\bar{\nu}$ is assumed to be the same as in ν, and J_{51} is of the order 10. The accreted material is roughly at a distance $r_7 \cong 1$. It takes about 4 MeV per nucleon to heat the material from $T = 2$ to 4 MeV. So, using (37), and taking $J_{51} = 1$, it will take about 0.1 sec to accomplish this heating. This is the correct order of magnitude.

The processes (36) use only a small fraction of the energy in the neutrinos, because of the low cross section. But they are, at least in principle, the mechanism suggested many years ago by Stirling Colgate, i.e., that the neutrinos from the inside heat the outside. Once heated, the material expands and exerts pressure on the shock. Probably this mechanism works only if the shock *almost* works without it.

In a second computation, Wilson showed that such a delayed resumption of the shock can also occur in the star of 25 $M_\odot$, using the new pre-supernova conditions of Weaver et al. described earlier. In other words, one should not give up hope when the shock becomes an accretion shock. Cooperstein's unsuccessful shock, if pursued to late times, will probably also become successful.

Non-spherical Stars

Bodenheimer and Woosley [16] have investigated the behavior of stars which are initially rotating at a reasonably high speed. The Fe core collapses in essentially the same way as has been described. It is assumed that the shock produced in this central core gets stuck and does not eject material.

However, the material somewhat farther out will now be subject to considerable centrifugal forces. Material falling in along the rotation axis will still accumulate onto the central core, but material coming in near the equatorial plane will be turned around by the centrifugal force and will be ejected. Woosley et al. estimate that about .5 $M_\odot$ can be ejected in this way, with an energy of the order .5 foe. This is certainly an interesting possibility, especially because many massive stars have substantial rotation.

This asymmetric supernova model will not leave a neutron star as a remnant, but a black hole. The mass of the central star which remains behind is considerably greater than the stability limit of neutron stars of about 2 $M_\odot$. According to observations and to statistics, at least a large fraction of supernovae leads to neutron stars. Therefore the rotating models of Weaver et al., while interesting as an existence proof, cannot explain all supernovae.

Summary

After many years of theoretical study, we now know three models which may lead to a successful supernova explosion:

1. Very low initial entropy, as calculated by Cooperstein [12]. This leads in a straightforward manner to a successful shock, which gets out to the surface of the Fe core in a time of the order .01 second. A neutron star is left behind, of mass around observed mass of 1.4 $M_\odot$.

2. A delayed shock as calculated by Wilson. The shock stalls for a long time in the mine field, but gets going again after a time of the order .2–.5 second. The mechanism involves low density outside the shock and some help from the neutrinos which stream out from the inner core. This mechanism also leaves a neutron star, perhaps of somewhat greater mass than mechanism (1).

3. Supernovae produced in a rapidly rotating star, as calculated by Bodenheimer and Woosley. In this model, the material is expelled from the supernova along the equator. The remnant will presumably be a black hole.

Acknowledgments

Much of the material in this paper is part of a forthcoming paper by G.E. Brown, J. Cooperstein, and me. I am obliged to both of them for permission to use this material ahead of publication. I also want to thank the other authors mentioned in the Introduction for their constant help in this research.

References

[1] Weaver, T.A., Zimmerman, G., and Woosley, S.E., (1978) *Ap. J.* **225**, 1021.

[2] Woosley, S.E., Weaver, T.A., and Fuller, G., (1983) *private communication.*

[3] Arnett, W.D., (1977) *Ap. J.* **218**, 815.

[4] Goldreich, P., and Weber, S., (1980) *Ap. J.* **238**, 991.

[5] Yahil, A., and Lattimer, J.M., in *Supernova, A Survey of Current Research*, Rees, M.J., and Stoneham, R.J., eds. (D. Reidel, Dordrecht, Holland, 1982).

[6] Bethe, H.A., Brown, G.E., Applegate, J., and Lattimer, J.M., (1979) *Nucl. Phys.* **A324**, 487.

[7] Fuller, G., (1982) *Ap. J.* **252**, 741.

[8] Cooperstein, J., 1982 *thesis*, SUNY at Stony Brook.

[9] Lamb, D.Q., Lattimer, J.M., Pethick, C. J., and Ravenhall, D.J., (1978) *Phys. Rev. Lett.* **41**, 1623; (1981) *Nucl. Phys.* **A360**, 459.

[10] Bethe, H.A., Brown, G.E., Cooperstein, J., and Wilson, J.R., (1983) to be published in *Nucl. Phys. A.*

[11] Yahil, A., *private communication.*

[12] Cooperstein, J., Bethe, H.A., and Brown, G.E., (1983) *unpublished.*

[13] Brown, G.E., Bethe, H.A., and Baym, Gordon, (1982) *Nucl. Phys.* **A375**, 481.

[14] Burrows, A., Mazurek, T.J., and Lattimer, J.M., (1981) *Ap. J.* **251**, 325.

[15] Wilson, J.R., (1980) Ninth Texas Symp. on Relativistic Astrophysics, *Proc. NY Acad. Sci.* **336**, 358.

[16] Bodenheimer, P., and Woosley, S.E., (1983) *Ap. J.*

19 How Well We Meant

I. I. Rabi

The third time Nick Metropolis called to ask me about the subject of my paper for this volume, I knew I had to give an answer. I asked him to wait ten seconds, then told him, "How Well We Meant"—a very nice title because we all know what is meant by that phrase, but we don't know the same thing. So writing this paper reminds me of Niels Bohr, who would start a paper by sending off something to the journal. Then, when the galleys arrived, he started writing the paper. There was the time he was trying to work with Dirac on some sentences, and Dirac finally had to say, "My mother taught me not to start a sentence before I knew the end." Well, I intend to do something of this sort by expanding on the title, "How Well We Meant" and by describing the situation forty years ago.

First, I want to back up and give you some idea of the backgrounds of the people who started Los Alamos. History is very important (although, unfortunately, it is being neglected now), because from the perusal of history you see the greatness and the folly of humanity. What I come to later in my description will be just that: the greatness and the folly.

It is important to know something about what this country was at the time Los Alamos started. This country—in fact the whole two western continents—is a land of newcomers. This started 50,000 years ago when

NEW DIRECTIONS IN PHYSICS
The Los Alamos 40th Anniversary Volume
ISBN 0-12-492155-8

there was a land migration from Asia. Right now we are on land that was originally occupied by this migration. Then, 500 years ago, people began coming over by ship. When these peoples came, they brought themselves and their skills, habits, and customs—all of which gradually blended.

In the course of time, the immigration that came by sea from Europe and Africa built here a most remarkable culture. As President Lincoln expressed it, "[it was] conceived in liberty and dedicated to the proposition that all men are created equal". These people made a unique country. There was no other democracy, no other country like this country. It was made artificially with a constitution that was made artificially. Usually such things are not very stable, but this is probably the most stable government that exists on earth, and, at the same time, fundamentally completely democratic.

This takes us up to the year 1900. I like the year 1900 because that's when *I* joined the oceanic migration and arrived in New York—the great city of the United States, you all agree. It was a most exciting time, because a few years before my arrival, the electron, x-rays, and radioactivity were discovered. Notice that all these discoveries were made in Europe, even though the United States was a free country and approximately 150 years old. The first president of the American Physical Society, Henry Rowland, noted in his inaugural address that if you looked back at the history of physics in the United States for 200 years, you could find only two scientists of any significance: Benjamin Franklin and Joseph Henry. So, although this was fertile soil for the development of a wonderful political institution, it was not so great for physics.

Well, the history of physics went on with the invention of Planck's constant, light quanta, Einstein's relativity, and Bohr's model of the atom. In 1916, I was a graduate of the New York City public schools, which were excellent, and a graduate of that great university in New York State, Cornell, but I had never come across the quantum. So you see that, although Willard Gibbs was flourishing somewhere in the latter part of the century and Michelson and Morley were having fun with light, we were not a great basic scientific country. In fact, I can easily illustrate our scientific standing in Germany. When I visited the University of Göttingen in 1927, I learned that they subscribed to *Physical Review*, but they had the twelve issues sent at once at the end of the year to save postage.

In 1927, I left for Europe to go to the wellsprings of physics. It's not that I went to learn the literature—that we knew, that we studied—but to learn something that I find very difficult to describe. To put it simply:

I knew the score, but not the melody. What I went to learn can be transferred by people and perhaps only by people.

We are now approaching 1943. In Europe I met Oppenheimer, Loomis, Stratton, Condon, Robertson, and many others. We had a common bond in that we were all treated by the Europeans, principally by the Germans, most condescendingly. And their attitude was proper; we were a great big country, rich in population—almost 200 million—and we hadn't yet contributed to physics in the same way that they had.

Well, we came back—all approximately in 1929. As I remember, the Washington meeting of the American Physical Society was on the lawn of the Bureau of Standards. We could get everyone into one good size room. Ten years later there was a war. How did we get the people to man the war laboratories, such as the Radiation Lab in Cambridge and all the labs of the Manhattan District? We had produced a seemingly infinite number of highly trained people in those ten years. Also, by 1939 *Physical Review* had become the leading journal in the world. You see how a culture can change.

How were these scientists produced? I believe it was the people in my generation—a Van Vleck in Wisconsin, a Lawrence and an Oppenheimer in Berkeley, the people at Harvard, the group in New York—who were able to produce such highly gifted, highly educated physicists in such a short period of time. These new people were the ones who made Los Alamos. Moreover, they—and I—were educated and trained without government intervention. For most cases there was no government money. When I got my degree from Columbia in 1927, there was no worry about a job. There were no jobs—a simple solution. So the training was done somehow or other by a great intellectual, perhaps spiritual, impulse.

Thus we came of age about the time we went to war and started these various laboratories. There is no question in my mind that we had at Los Alamos one of the great assemblages of people of highest talent. Most were young, were inexperienced in the task they were set to do, and were starting from scratch. The same was true of the other big laboratory, the Radiation Laboratory in Cambridge. Let me give you some flavor of how fast the problem could be attacked even though we were starting from scratch. After the discovery of Hahn and Strassmann and the interpretation of the effect of fission, Fermi was able to calculate the size of the crater that one kilogram would produce. So we had the men.

There is one other important thing you should know about the background of the people who accomplished these very great things at Los

Alamos. Store-bought equipment was rare. For experiments you made your own equipment. A lot of skill had been developed in the early days of the automobile industry repairing your own car. I know Nordsieck spent a lot of his time keeping his car repaired. So, in addition to the theory, we were accustomed to hands-on physics.

Forty years ago, Bob Bacher and I came up together to Los Alamos to attend the session whose anniversary this volume commemorates. Neither of us knew very much about what was going on. We had been met by Eugene Wigner in Chicago, and he was supposed to have given us some information about the project. He warned us not to spend too much time on the inefficient bomb but to work on the efficient reaction directly. We didn't know quite what he meant. Later, we figured it out: Don't waste your time on the fission bomb; go directly to the thermonuclear.

The day we came up to this wonderful place happened to be one of those spring days when everything was lovely. The air was clear and mild, the Sangre de Cristo Mountains were distinct and sharp, the mesa on the other side beckoned. The ride up the old road and over the old bridge was somewhat hair-raising but very interesting. Then, of course, there were the Indians. We certainly seemed to enter a new world, a mystic world.

Now what about the people who were here? First, Robert Oppenheimer—of all people to select as director. It was astonishing! He could drive a car with only occasional accidents but never fix it. But he was a man of brilliant insight with a command of language that was very elevating. He set a high-level tone. Then there was General Groves, his boss, whom most of the scientists who worked at Los Alamos remember as a born malaprop. Now, that combination of Oppenheimer and Groves was remarkable because you would always tend to underestimate Groves, but he was the power behind Oppenheimer. That combination made the thing work. As for the other people, somebody remarked, "Look at the list of who was here then. It was a remarkable group. Not everybody got a Nobel Prize, but many of them did."

Also, the job got itself organized in the most remarkable way. I don't know just how that happened—a sort of spontaneous production in which we organized around the various parts of the project. Especially important were the theorists. The project couldn't have been accomplished without great theoretical insight because, at the beginning, there were no materials for the bomb that we could assemble, and their properties were not known.

One criticism of the people who worked at Los Alamos is that, in view of what we were working on, it was somehow immoral to have had such a good time. And we did have a great time, because we were young and full of spirits. But we worked very hard at the central problem. It was hard to divert anyone to have any real fun. For instance, I tried to have some fun when Paris was recaptured. I happened to be on the Hill at that time, and I said, "We ought to celebrate this event in some way." So we organized a small parade—maybe four of us—and we walked down the road singing the Marseillaise. Nobody followed. I said, "We'll go to Stan Ulam's house because Françoise is French, and we'll serenade them." It didn't work. She was not well. What we finally did was what happened at all attempts at celebration. We went to the Serbers where Charlotte was ready with drinks, pastrami, and Jewish dill pickles. This was how we celebrated the recapture of Paris.

We did have one great party—one that I'll never forget and that the people who were there will never forget. It was the party that Ken Bainbridge and others threw when I got the Nobel Prize. It was well worth the effort.

When the Germans surrendered, the powers that be (I don't know where they're situated but probably in some cell in Washington) decided that this was no time for a celebration. They felt that the real war—the one with Japan—was yet to come. And actually there was no celebration. However, I had been carrying around two bottles of Johnny Walker Black Label, one dedicated to the fall of Hitler and the other to the fall of Tojo. So we did celebrate slightly in Hitler's case, but surreptitiously.

It's inappropriate for me to go into the work at Los Alamos that had to be undertaken to develop the weapon, to develop the various experimental and fabrication techniques, and even to develop the industry. In fact, that industry had to go into something that had never been done before, using materials that had never been worked on. There are other people who know more about this. I will go on to the final test.

As Trinity approached, tension became greater and greater. Ken Bainbridge, who ran the test, did a marvelous job organizing it. Oppenheimer invited me to be present, and I came to Los Alamos. We drove down together to the test site located in a very romantic desert—Jornada del Muerto. The trouble was that when we got to this real desert where the test site would be, it had the only green things around that we could see. "What sort of desert is this, anyway?" "Oh, you're a New York boy. What do you know about deserts?" Well, that afternoon it started to rain. This desert rain was really something. Anyone who was there can remember the mud and walking around with clods of earth clinging to your feet.

Not having a role in the test, I joined the meteorologists because I realized that the decision of whether or not the test would take place did not rest with the scientists, but with the meteorologists. They wanted the fallout on the site, not all over the place.

Well, there it was: the bomb, or gadget, up on the tower, thunder and lightning all around, the rain falling heavily. We had meteorological information from the North Pole, the South Pole, Japan, Russia—the best information to be gotten at this time—that said it was supposed to be clear with visibility up to sixty miles. But what it looked like we'd have instead when the thing was to go off was a tremendous storm. Well, what to do? We played poker! Then the word came down that it would be possible around 5 o'clock in the morning.

Indeed, at 5 o'clock the sky was clear. I was near the main station where Conant and other people were. Enrico Fermi was to one side ready to do his famous experiment. Then the gadget went off. People with greater poetic gifts than mine have described the situation, the effects, and what they saw. I still remember, a few seconds after it went off, Enrico dropping bits of paper to measure the strength of the blast.

I was tremendously impressed by this bomb, by the sights. I'm not going to try to describe it, but the effect was visceral. It really penetrated. For a few minutes, I was very pleased at the success of the effort. Then I began to have gooseflesh because I realized it was the end of one world and the beginning of another—a much more dangerous world than what had gone before. We now had a power that hadn't existed and that put humanity on a new plane of power and responsibility. It was not days afterward, but almost minutes afterward, that those who had some feeling for it realized they'd witnessed an event that changed their lives, that put upon them a responsibility that hadn't existed before.

I think this feeling, more or less, permeated through the people in the Laboratory who had direct, or even indirect, connections with the effort. Something irreversible had happened. The knowledge became public. By that I mean more than two people had it. We would have to live and deal with it from then on. Robert Oppenheimer and I talked about it constantly. What to do? At that time only we very few—the people at Los Alamos and some others—knew about and really understood this thing. The bomb and its effects were just words to others. We who had been there and realized what this was knew we had something to dedicate ourselves to.

It was not an entirely new sensation. I had something of the same kind in 1940 when I went to help start the Radiation Laboratory in Cambridge and began to have serious contact with these military problems. It made me realize how vulnerable people are to things that science can invent that manipulate, that change, that drive, that kill people. With the nuclear weapon, we could see very easily the developments that would come from it.

The afternoon of Christmas after the test, Robert Oppenheimer and I laid down the outline for what later became the Baruch plan presented to the United Nations in 1946. To this day I look back in disbelief at the official United States position then. All you have to do is read the newspapers now and think of what the United States presented at that time, that is, of giving up the atomic bomb and putting it under international control. Moreover, we realized that proliferation was bound to occur. We knew we had to put the bomb under controls that would be self-enforcing. I doubt that there will ever really again be new ideas in kind. If you look back at the newspapers of that particular period when the bomb was nascent, you'll see that the general public understood this new thing and its nature much better than the public understands it now. The people then realized the stages possible for nuclear weapons and were very much alive to it. As a result, you *could* have a proposal of that nature from the United States.

I started by mentioning human greatness and human folly. We gathered on April 14, 1943, to see what we could contribute toward meeting the problems confronting civilization. We were in such a terrible state then, both in Europe and the Pacific. The war had been going very, very badly. There were glimmers of hope, but almost every effort we made on both sides—the British and ourselves—was unsuccessful. The Nazis had overrun Russia, and we were just being pushed out of North Africa. There was every incentive. It was not just patriotism for the United States but patriotism for western civilization, for the ideals of humanity in which we had been brought up. The world seemed about to be engulfed by a fanatic, barbarian culture. And it did look then, as it turned out to be, that only through science and its products could western civilization be saved.

Well, we saved it. I say this proudly, and, I think, truly. We designed and made the weapon in Los Alamos, then handed it over to the delivery boys. Having given this great power to our country, we were in a position to start on a new road toward a new world. What did happen? At first, the military realized this was a new deal and all the old rules were off. But it didn't take them long to say, "Yes, we have this bomb. Now we have to protect it." At that time, I thought naively—no, sensibly—that we would need very few weapons to match the explosive power that we'd used in the whole war on Japan and Europe. But no!—we had to protect it, we had to have a supply of them, we had to develop them further. It evolved into a race for itself, in itself.

In the postwar period, we forgot the basic reason the United States entered the war. During the war, we had been damaged when the Japanese attacked Pearl Harbor, but otherwise we were quite safe. We had entered the war to save and protect civilization. That was the purpose, that was the motive power of our effort—to protect civilization and, in the process, to save the United States. But the way things developed—and this is the folly—it became a thing in itself. The question now is not so much how to protect civilization, but how to destroy another culture, how to destroy other human beings. We have lost sight of the basic tenets of all religions—that a human being is a wonderful thing. We talk as if humans were matter.

My purpose in this paper is to bring back the original human orientation to our perceptions. There is no way for scientists to escape the responsibilities of their knowledge. Our experience with the bomb taught us what happens when we hand over the products of our knowledge to people who don't have that knowledge or even a fundamental appreciation of it, to people who don't have a feeling for the glory of the human spirit, to people who don't respect science as the highest achievement of mankind and as the process that takes us out of ourselves to view both the universe and our place within it.

We now have the nations lined up like those prisoners at Auschwitz, going into the ovens and waiting for the ovens to be perfected, made more efficient. I submit that this fatalistic attitude is very un-American. It is not American to stand around waiting for something to happen, hoping it won't, when you see it on the horizon. It would be much more true to our spirit to understand and prevent it. The United States was founded on a very revolutionary principle. It was the example of the United States that toppled most of the kingdoms and empires of the world. We did this by asserting the greatness of the human spirit. Somehow, rather than this calculus of destruction, we must get back to our true nature as a nation and as part of western civilization.

So I return to my title. Although we meant well, we abdicated. We gave it away. We gave the power to people who didn't understand it and who were not grown up and responsible enough to realize what they had. We always hoped we would elect better, broader people.

We've talked about arms reduction and arms control now for thirty years. I've listened to this talk inside and outside government. What is the result of all this talk? Well, we don't know what the result is. What we did at Los Alamos was great, what we did was inevitable, what we did was fortunate for the United States and for the world up through the period just after the war in 1946. The submission of the Baruch Plan was to have been a step forward in human organization, but the process got taken out of our hands. How do we recover it? We cannot put this evil spirit back into the bottle. We have to learn to live with it. Part of our responsibility as scientists is to find ways, devices, connections to learn to live with it. We had a partial attempt in 1955 at the Conference on the Peaceful Uses of Atomic Energy. That was one place where we got together with the Soviet scientists and found them to be wonderfully like ourselves. On that level, the supernational level, there were no problems, or no very great problems. It's at the level of culture that our real problems come. Our culture is less and less touched fundamentally by science, and our leaders are selected from such cadres within our culture. We must attack this problem if we are to reinfuse our culture with its original democratic spirit. That's the challenge.

20 History and the Hierarchy of Structure

Cyril Stanley Smith

Partly because of my intimate involvement with the many phase changes in metallic plutonium and partly because I have long been puzzled by the differences between the physicist's and the metallurgist's views of structure, I have become interested in the nature of change, i.e., of history, whether in material systems or in the conceptual patterns in the human mind. And change involves choice.

Some Notes on the History of Solid State Physics

Not much has been published on the history of the science of materials as distinguished from that of matter: Interest has focused on concepts of atom and cosmos, with neglect of structure on the human scale. I have come to believe that aesthetic curiosity precedes discovery of relationships, whether in purely intellectual fields or in technology. Until the present century, art seems usually to have preceded technology, and problems for the philosopher were suggested by observations in the studio or workshop. Understanding itself seems to be primarily sensual, but it is not necessarily correct, and resort to logic or experiment is essential to test its propositions and to enable them to be communicated or used.

NEW DIRECTIONS IN PHYSICS
The Los Alamos 40th Anniversary Volume
ISBN 0-12-492155-8

The first ideas on the nature of matter that were expressed by the Greek philosophers arose out of the sensual experience of the artisan. The "atomism" of Democritus and Lucretius is based on what a smith or sculptor sees as he works with his materials and relates texture to how they behave. Similarly, Aristotle's four elements and elementary qualities, which are laughed at by most of us today, classify the variable workability of materials in terms of the qualities associated with the three states of matter (solid, liquid, and gas) and the role of energy in modifying them. The later salt–sulphur–mercury principles of Paracelsus are clearly based on sensual recognition of the different properties associated with what today we call ionic, Van der Waals, and metallic bonding. Even the alchemists' mystic diagrams were based on a recognition of the subtle relationship between the hierarchy of connections perceptible to the human eye and brain and those underlying the multitudinous properties of matter.

In the seventeenth century atomism was revived in the lively corpuscular theories, especially those of René Descartes and his followers, who advanced some remarkable intuitive speculations on the ways in which the properties of matter depended on the shapes, arrangement, and orientation of parts, the ultimate indivisibility of which was denied. Energy became a particle, called caloric, which was fixed or free depending on its association with the other parts. Perhaps most interestingly of all was phlogiston, that almost but not quite intangible particle, the presence or absence of which was supposed to account for the particular properties associated with metals. Caloric can be regarded as a precognition of Planck's quantized unit of action, while today's electron, as it moves between the energy states within and between atoms, is very close to being what the phlogistonists imagined.[1]

All these are ideas central to the solid state physicists' domain today, but it was craftsmen and chemists who first formulated and made use of them. However, the concepts did not lend themselves to mathematical expression, and, following the great Newton, science had become so heavily mathematical that Cartesian corpuscular speculation was discouraged and despised. (Dare one suggest that the introduction of the calculus with its great power actually delayed the advance of some branches of science? Fourier analysis and quantum theory work so well because

---

[1] See Chapter 5 in the author's book, *A Search for Structure*, MIT Press, 1981. Also "The Prehistory of Solid State Physics", in *Solid State Physics, Past, Present and Predicted*, edited by Dennis Weaire and Colin Windsor, Hilger, 1987. This emphasizes the importance of medieval corpuscular philosophy and of the structural concepts of Descartes and Emanuel Swedenborg.

they depend upon binary interaction and in no way require infinitesimal or sinusoidal substructure.)

Though in the eighteenth century particles were generally taken for granted, they could not be quantitatively treated prior to Dalton's atomic combinations. Thereafter, the structural chemistry of the molecule flourished, but the physics of real structures was limited to those few cases that involved only homogeneous elastic behavior. It was well over two centuries after Descartes' imaginative beginning before solid state physics revived. When it did, it depended heavily on the mass of observations and rather naive theoretical models that had been gathered by metallurgists.

For many centuries physics was totally unable to handle the most interesting properties of matter—the structure-sensitive ones—but from the 1860s on metallurgists, especially those working on that most complicated alloy, steel, began to show how structural features on the scale readily accessible to the optical microscope could be observed and manipulated with profound effect.

The history of microscopic metallurgy is a kind of prehistory of solid state physics.[2] Then came the discovery of X-ray diffraction and the possibility of determining averaged lattice structures with considerable precision, soon followed by the realization that structural imperfections were at least as important as the ideal symmetries that had been earlier sought. The concepts of the lattice dislocation and of Schottky and Frankel defects mark, I believe, the beginning of a fundamental change in the nature and philosophy of science: Added to the uncertainty at the quantum level there is now the more general recognition of the influence of unpredictable history in establishing local structures, and hence of the necessity to know about actual historically generated structural peculiarities on all scales if the behavior of material aggregates is to be understood.

During the last four centuries virtually every advance of science has involved the postulation and discovery of new levels of structure followed by mathematical analysis of their relationships within exactly defined boundaries. Today the boundary is seen to be not the limit but the very problem itself. What lies within it, however necessary, is repetitious and of no more than statistical interest. The scale of resolution used to detect differences in structure between the two sides of the boundary is purely a matter of human choice, and it must be recognized that the boundary in actuality is not in the least like a pencil mark on paper and even less like the mathematician's idealized one. Boundaries are chains or epidermal

---

[2] See C.S. Smith, *A History of Metallography*, University of Chicago Press, 1960. Paperback reprint, MIT Press 1987.

layers of granules or corpuscles forming a substructure of whatever scale, dimensionality, and complexity is needed to form a closed transitional region between the more directionally balanced substructures in the two regions that the boundary separates.

On Aggregation and Segregation: The Hierarchy of Choice

Although the truth of a concept can only be established by logical analysis or experiment, and its application is easiest when it is expressed in mathematical form, the *generation* of a concept in the human mind is not logical. Understanding begins with a stage of aesthetic metaphor prior to that of logical analysis. It is easier for most people to understand the nature of relationships when shown in diagrams than when expressed in mathematical equations. We seem to be able to sense topological relationships more easily than metric ones—witness the fact that we perceive the unchanging identity of an object despite changes in measurable angles and distances as our viewpoint changes. The topology of two-dimensional diagrams is therefore of particular interest.

Connections of higher dimensionality than two are correctly represented in a two-dimensional diagram by the number of lines intersecting at a vertex, i.e., by what will be called their valence. The distinction between a point, a polygon, and a polyhedron is entirely an internal affair and progressively appears or disappears as the scale of resolution is changed. The higher terms in the Euler–Schäfli–Poincaré relations are unnecessary to the analysis of continuity and discontinuity of connections. The 2-D form is

$$P + V = E + 1, \qquad (1)$$

where P is the number of closures (polygons), V the number of vertices, and E the number of lines connecting them.

Each line has two "ends", and each vertex of valence r has r lines meeting at it. Since each line is shared by two polygons except for those lines at the outer boundary of the array or those that terminate in a monovalent vertex (denoted by E_b and E_0 respectively),

$$2E = \sum rV_r = \sum nP_n + 2E_0 + E_b, \qquad (2)$$

where V_r and P_n are, respectively, the numbers of vertices of valence r and of polygons having n sides. Substituting the sum of these two values for E in Equation 1 and simplifying gives

$$\sum(4 - n)P_n + \sum(4 - r)V_r = 2E_0 + E_b + 4. \qquad (3)$$

Islamic mosaic patterns (Fig. 20.1A) beautifully illustrate the interdependence of vertex and polygon. The centrality of the number 4 is well illustrated by the fact that in an array consisting of quadrivalent vertices, the sum $\Sigma(4 - n)P_n$ always equals $E_b + 4$. Moreover, if the dual of any 2-D net is superimposed on it and the two joined to create quadrivalent vertices wherever they cross, the entire array is reduced to tetragons (Figs. 20.2A and B), and, in any array of tetragons such as the Penrose tiling in Figure 20.1B, the entire balance of the structure is represented by the averaging of the non-quadrivalent vertices. Note in Figure 20.2A how departures from 4 are compensated in quincunxial array and also that chains of tetragons form boundaries between regions of high and low density.

If the hierarchy of choice is represented, as it can be, by nets using only trivalent vertices, then $2E = 3V = 6(P - 1)$ and $P = V/2 + 1$. The polygon sides conform to the following:

$$\sum(6 - n)P_n = E_b + 6 \tag{4}$$

These equations, which were given in Chapter 1 of the author's book, *A Search for Structure* (1981), are quite universal and give a succinct description of the hierarchy of connections in any system. Everything

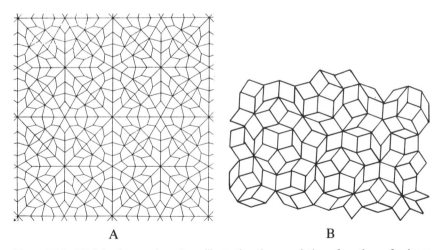

A   B

Figure 20.1. (A) Islamic mosaic pattern illustrating the association of vertices of valence 3, 4, 6, and 8 in accord with Equation 1. (From J. Bourgoin, *Arabic pattern and design*, Dover, New York, 1973.) (B) An example of non-periodic tiling after Roger Penrose, based on quadrilateral polygons meeting at vertices having valencies between 3 and 7 in ratios to yield, internally, the average of exactly 4.

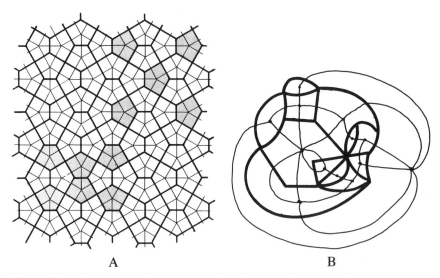

A   B

Figure 20.2. Diagrams showing the generation of exclusively four-sided polygons by the superposition and joining of a 2-D net and its dual. (A) Excised portion of a lattice of pentagons. (B) A distorted heterogeneous net. Note that any 2-D net can be considered as a Schlegel projection of a map on a sphere, and the environment provides a single vertex for the dual net: If monogons are excluded, E_b for the composite is always 4.

within a chosen boundary, whether regarded as polygons or vertices, joins to give the total $2E$, and every external connection is acknowledged but is not further examined. However, in common with any statement that is universally true, these equations are useless until some basis of distinction is defined, in this case necessitating a distinction between a closed vertex and an open polygon. This requires a definition of scale and a means of observation. Ultimately this involves only the ability to choose between radial and circumferential directions in the garden of forking paths (Shades of Borges!). This is why the two-dimensional bubble model with only trivalent vertices is emphasized in what follows. It represents the most basic combination of physical, chemical, mathematical and aesthetic factors: Practically every stage of the author's development since childhood has been enriched by playing with it.

Figures 20.3 and 20.4 show two 2-D arrays of bubbles. In the first there is a small volume of liquid under negative pressure, and the bubble size is far from uniform. Surface tension requires that the films meet in equilibrium at 120 degrees, and the pressure differences require a distribution of the number of sides of the resulting polygons around the average of six in accord with Equation (4). The bubbles in Figure 20.4

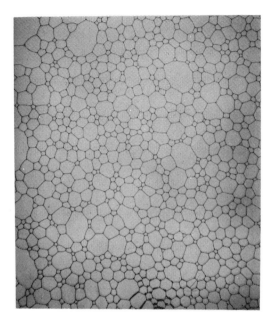

Figure 20.3. An array of soap bubbles in an essentially two-dimensional cell. Reduced by two-thirds.

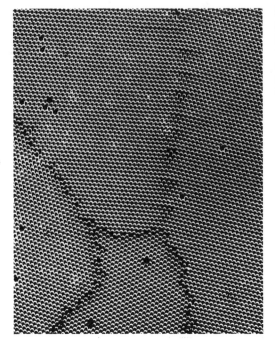

Figure 20.4. Two-dimensional array of bubbles ca. 0.3 mm diameter illustrating the formation of regions with hexagonal close packing and the formation of grain boundaries, dislocations, vacancies, and interstitial defects.

are extremely small, floating as a raft on a horizontal liquid surface. They are both mobile enough to achieve close packing except where regional strain intervenes and compressible enough to accommodate to both vacancies and directional misfits within rather small distances. This is the well-known Bragg–Nye model of crystal imperfections. Many early models of crystals were based on the stacking of hard balls. In the 2-D version of Figure 20.5, all contacts are quadrivalent and the rigidity of the stack sometimes results in the transmission of the strain of local misfit and disturbance of the topology over considerable distances.

On Quantity, Quality, Substructure, and Color

The submergence of trivalent vertices within an epidermal layer of cells suggests a method of separating quantitative and qualitative aspects of a structure. The average vertex valence, exclusive of those vertices on the boundary, is not affected by the actual distribution and is a quality akin to density, while the sum of interior valencies plus that of the boundary vertices is akin to mass.

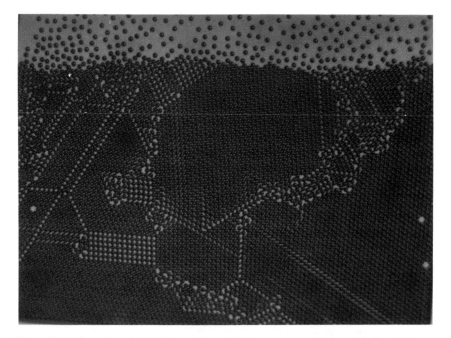

Figure 20.5. Crystal model similar to Figure 20.4 but composed of steel balls, 0.85 mm diameter.

The transition from the level of substructure to that of macrostructure marks the distinction between quantity and quality. The density of matter is independent of its mass and environment, just as the color used to designate a region on a map is independent of the size of the region. However, note that this ceases to be true when the scale of epidermal granularity becomes significant—as gravitation in the former case and as human perception of color in the latter.

All structure is hierarchical and cellular. An intensive quality such as color attributable to extensive macroscopically bounded regions depends upon the presence of substructure that is unresolved but becomes resolvable if the scale of resolution is changed. It is a topological property, not a geometric one, related to the density of lines internally associated with, but not affecting the boundary of, each macro region. This suggests an approach to the old four-color map problem. (The reader is warned, however, that this has not been critically examined by mathematicians familiar with the problem).

In any macro net composed of $4N$ polygons, the addition of $6N$ substructural lines without change in the number of vertices will generate $6N$ additional polygons. Equation 1 relates the numbers of polygons, vertices and lines in a simply-connected 2-D net under all conditions. Using the subscript c to denote the values in the colored net, $P_c = 5P/2$, and $V_c = V$. The most difficult map in which to establish color contrast is one with only trivalent vertices. Vertices of higher valence, r, may be dissociated to yield $\Sigma(r - 2)V_r$ additional trivalent vertices and $\Sigma(r - 3)V_r$ additional lines without changing the number of polygons to be distinguished.

In a map in which all vertices are trivalent, $P = (V + 2)/2$. After "coloring", $P_c = V/2 + 7$, and $E_c = E + 3P/2$. The additional lines may be distributed in such a way that each member in each and every group of four contiguous macro polygons has a substructure that can be distinguished from the substructure of all its neighbors by the internal presence of 3, 2, 1, or 0 substructural polygons.

Though it is time-consuming to find the correct distribution to apply to a pre-existing net, it is possible to generate a colored map of any complexity by assembling, with successive choice of size and orientation, any number of groups of four macro polygons, as in figure 20.6, each having four trivalent vertices on the outer boundary, and two inner vertices, one of which is trivalent and lies on the boundary of a cell with no internal color, while the other (which has valence 15) is shared by six "color-giving" polygons that can be distributed between the three macro cells in a manner to give the three different levels of density needed to distinguish them.

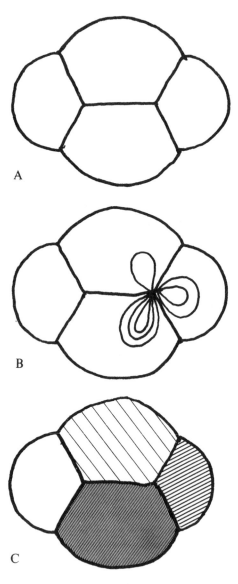

A

B

C

Figure 20.6. (A) Four polygons joined at 6 trivalent vertices. (B) Distinction between the densities of their sub-structures achieved by the addition of 6 lines defining six internal polygons. (C) The same distinctions represented by gross density of shading in four different levels.

Contact between the 4-groups generates additional trivalent vertices and lines, but does not change the number of polygons as long as entrapment is avoided.[3] Figure 20.7 shows an assembly of 32 macro polygons with

---

[3] If additional polygons are formed, as by the junction of a 3- or 4-color ribbon against an outer wall of cells, new vertices are generated and the numerical balance is unaffected; however the color distribution among the neighbors may need to be interchanged. Note

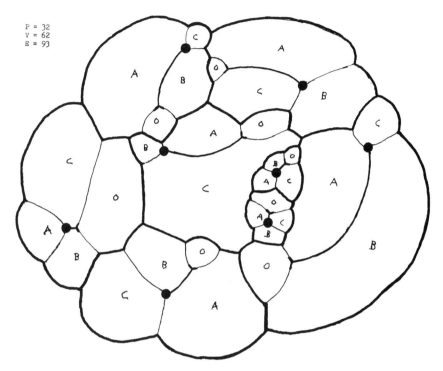

Figure 20.7. An irregular net of 32 polygons formed by 93 lines meeting at 62 trivalent vertices. Subgroups of 4 polygons are outlined and interval vertices serving as color centers are emphasized. The net is a distorted assembly of polygons as Figure 20.6B with the central pair of vertices and the associated colors oriented to maintain neighborly mismatch.

the boundaries of the groups of four and their "color centers" emphasized. Figure 20.8 shows the same array shaded as in a conventional geographer's map. Being topological, the relations are not affected by distortion; and since the contrast is in unitary steps it is not affected by any background color of uniform density superimposed on all regions. The same principles apply to the surface of a polyhedron or global map with trivalent vertices and $4N$ polygons.

When coloring a map in practice, one can start with any polygon (preferably one at the outer boundary that has many sides) and assign color O to it. Then, following along the ribbon of contiguous cells surrounding it, it is a simple matter to determine first the orientation of

---

also that since E alone is the measure of density, the same distinction can be achieved by any combination of substructural P and V, such as those in Figure 20.9, that yields $P_c + V_c = V + 5P/2$.

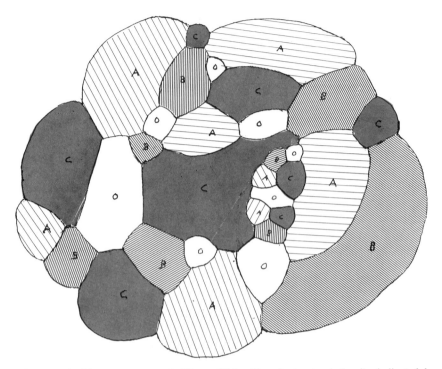

Figure 20.8. The same net as in Figure 20.7 with sub-structural density indicated by conventional shading.

lines joining paired trivalent and 15-valent vertices surrounded by four cells as in Figure 20.6B, and then the distribution of the remaining three colors, A, B, and C. This procedure is repeated with adjacent ribbons until the whole array is colored. The color of enclosed trigons is immediately obvious and an isolated monogon can have any color not that of its environment.

The only topological "truth" seems to be that $2E$ must always be equal to $\Sigma_r V_r$. This, regardless of dimensionality or genus provided only that every line is associated with at least one vertex. The stability of three-dimensional matter often seems to depend upon contrast of regional substructural densities around quadrivalent vertices.

On Dimension and Projection

The 2-D soap bubble aggregate with its trivalent vertices can accurately represent the division of space and hyperspace within systems of any dimensionality. We saw above how vertices of valence higher than three

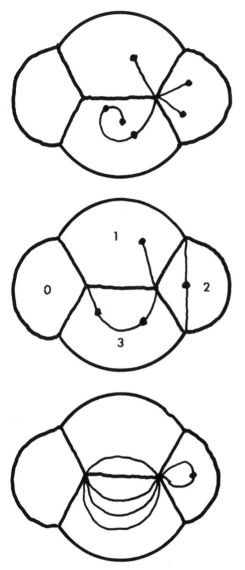

Figure 20.9. Alternative distributions of the lines that give the same local substructural density as Figure 20.6B without effect on the number of primary polygons or vertices in Figure 20.6A.

could be dissociated into trivalent vertices with the creation of additional lines but without change in the number of empty regions. The numerical density diagrammatically representable by lines and vertices, and which corresponds to events in matter can be connected or disconnected, compressed or expanded, to form nets of any dimensionality, but the emptiness without which they could not exist, can neither be constrained nor disconnected. The difference between a polygon and a polyhedron lies only in the perception of closures within the linear nets associated with them. A surface does not have two sides unless it has an unresolved material substructure.

The usual 2-D projection of a polyhedron superimposes edges from opposite sides, with the generation of "false" quadrivalent vertices (Fig. 20.10A.) Although this does not change the relationships expressed by Equations 3 and 4 or the continuity of connections, the clutter is undesirable and becomes intolerable in aggregates of many polyhedra. The Schlegel projection, in which one face of a polyhedron is opened to form the boundary of the 2-D net correctly represents all vertex valencies and connections, though not the number of polygons (Fig. 20.10B). Schlegel projections of cubes can be joined at their vertices in a quadrangular array to resurrect and incorporate the missing squares and the assembly then correctly represents all the features in a simple cubic lattice with its hexavalent vertices (Fig. 20.10C).

These unit pairs of filled and empty polygons not only represent the inside and outside views of the polyhedron but are also reminiscent of the way in which a practical foundryman makes two-part molds of sand in which to cast three-dimensional objects. He carefully plans the orientation of the model so that the surface dividing the mold can join the surface of the model along a closed line of maximum extent, following such

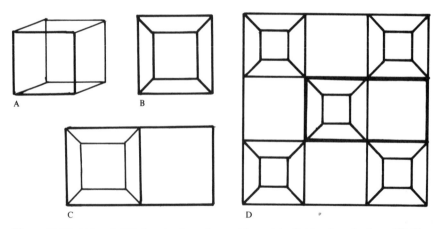

Figure 20.10. (A) and (B) A cube in orthographic and in Schlegel projection. (C) The Schlegel projection with the deleted polygon restored. (D) Assembly of Schlegel projections in quincunxial array. This combines the inner and outer views of a cube and generates the hexavalent vertices of the simple cubic lattice.

contours that there are no overhanging projections to prevent the withdrawal of the model from the mold. The simplest is the mold for a sphere, which is divided equatorially. A line with no vertex has been formed around the object. The mold, when open, forms two parts each with a hemispherical depression.

A foundryman wishing to cast a cube in a sand mold could divide the mold so as to make the parting plane continuous with one face, but it would be almost as difficult to withdraw the model without deforming the mold as it is to infer the structure of a cube from an orthogonal projection made normal to a face. A more natural division of a cubic mold cavity is that in which the cube diagonal is pressed into the mold and the parting surface is not plane but is curved up and down to meet the cube edges. In projection the two parts appear as hexagons, preserving two internal trivalent vertices and two on each part of the boundary but with two quadrivalent vertices at the "hinge" and four divalent ones that will merge and reconstitute trivalent ones on closure (Figs. 20.11A and B).

Shapes of more complicated structure can, of course, be represented by additional vertices within the cavities. Closure at the equator of the three dimensional mold cavity itself or by the one dimensional line surrounding the two joined figures in the projection correctly represents the

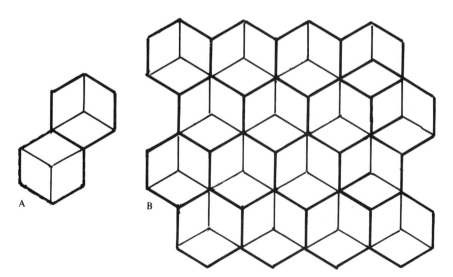

Figure 20.11. (A) The two halves of a foundryman's mold for casting a cube. (B) A pattern plate for forming a half mold for casting many cubes. (The surface is not plane.)

topology of the surface structure of the original object. The two parts, simultaneously joined and separated by the parting line, may represent difference in orientation, either in the plane of the projection or normal to it, and the connections from the internal vertices may be satisfied either at the boundary or internally to form some epidermal structure. Ultimately only *direction* matters. The designers of objects to be cast try to make them convex to simplify mold making, though the foundryman long ago learned how to attach inserts on the surface of the mold to produce complex concavities and to use cores to provide for internal cavities. In the projection, these become inserted polygons that do not join the outer boundaries or polygons with at least four sides that cross from one side of the boundary to the other.

The reconstructed Schlegel projection of Figure 20.10C (which might perhaps be called the Pandora projection) provides two-dimensional units that can be assembled in aggregates to represent any assembly of cells and the space within them. When regions join, the changed valence of internal vertices exactly represents the junctions of polyhedra on 3-D assembly.

Figures 20.12 and 20.13 show such projections of the polyhedra that constitute the space-filling units in the four non-simple cubic lattices and their assembly. Figures 20.14A and B are the units in stacks of triangular and hexagonal prisms respectively. The Schlegel projections of different polyhedra that combine to fill space can be assembled in 2-D nets in the

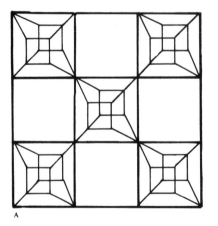 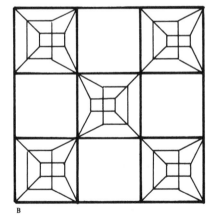

A   B

Figure 20.12. Schlegel projections of (A) rhombic dodecahedra, (B) trapezohedra assembled to generate the vertices in the face-centered cubic and hexagonally close-packed lattices respectively.

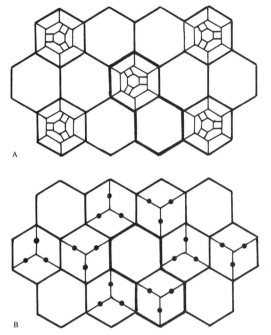

Figure 20.13. Assembled Schlegel projections of (A) truncated octahedra forming cells in the body-centered lattice, (B) the polyhedron with ten vertices and four (non-planar) hexagonal faces that stacks to form the diamond-cubic lattice. The latter is the only lattice in which the number of sides of a 3-D cell is the same as the valence of the vertices at which they join.

same way as above. Figures 20.14C, D, and E show this for the combination of tetrahedra and octahedra. Note that two shared polygons (triangles) have disappeared: The "real" basis of space filling at this stage has become a polyvertex with two trivalent and six pentavalent vertices. The subdivision of space by plane surfaces is a human concept, not an aspect of physical structure. In these diagrams the points representing atom centers in the usual crystallographers' treatment of symmetry have been replaced by linear boundaries representing both the separation of and the contact between regions of indivisible space. The topology is that of string theory.

The balance of density in systems of any complexity is represented by the distribution of vertex valencies to form the total $\Sigma_r V_r$ within boundaries of any chosen size. Since vertices of valence higher than 3 can be dissociated without changing "space", the nature of hierarchy can be modelled by a two dimensional array of bubbles and analyzed in two dimensions with the aid of equation 4. By focusing on connection, separation, and choice as represented by the assembly of trivalent vertices, the mystery that has surrounded spaces of dimensionality higher than

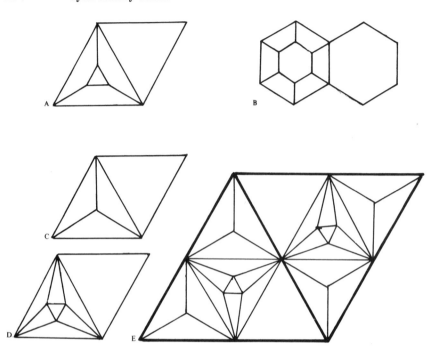

Figure 20.14. Schlegel projections of triangular and hexagonal prisms (A and B) which form, respectively, 3-D lattices of octavalent and pentavalent vertices. (C) and (D) are projections of tetrahedra and octahedra which associate in the ratio 2:1 to generate vertices of valence 12 (E).

two seems to vanish. The Schäfli–Poincaré relations between alternating positive and negative contributions of features of ascending dimensionality are replaced by simple sums of polygon sides and boundary vertex counts. It seems that the universe is sesqui-dimensional and that what we call higher dimensions are a manifestation of the unbalanced connections in the epidermal layer around regions of higher density, i.e., smaller bubble size! This approach has much in common with that of Benoit Mandelbrot in his book, *The Fractal Geometry of Nature* (W.H. Freeman, San Francisco, 1982), but the definition of dimensionality is simpler, and the hierarchy of closure as well as distinction is emphasized.

The replacement of any 2-D net by its dual simply inverts the connectivity of space and antispace; things that are connected and things that, for the time being, are inaccessible but not ignored. In the physical soap froth, the continuity of the liquid phase with its negative pressure extending through the whole system is balanced by the average of the positive pressures in the separate cells of the gas phase. A drastic change of resolution would permit one to observe, or rather represent, the structure as chains of epidermal cells with fewer than six sides (mainly four) surrounding and separating larger polygons, the structure of which can be ignored because it must be of average hexagonality. This suggests an approach to chemical thermodynamics based on the assembly of polygons in our trivalent net.

Substructural Thermodynamics

The symmetry around the number 4 that is expressed in equation 3 becomes an interesting dissymmetry when all vertices are trivalent and equation 4 applies. Polygons or groups of polygons within a fixed boundary then have a combining capacity somewhat analogous to chemical valence; this is expressed by our old friend, $\Sigma(6 - n)P_n$.

Obviously, polygons can have "valencies" of any magnitude, but these must be satisfied either at the external boundary (the E_b in equation 4) or internally by an appropriate larger number of polygons with less than six sides. A digon can sit on the boundary between any two polygons, but a trigon must make a definite, though local, choice between possible neighbors, and tetragons and pentagons associate with less freedom. The difference between internal boundaries and external ones is striking. The possibility of rearrangement within a local grouping depends directly on the "$(n - 6)$ness" of their assembly, which is reflected immediately by the boundary: No internal analysis is necessary. There is an analogy with the degrees of freedom expressed by the phase rule of Willard Gibbs.

A chemical entity can be represented by a grouping of polygons within an effectively inviolate boundary, the scale of which determines whether it is an electron, an atom, an ion, a radical, a molecule, or a phase. The aggregation of bubbles in arrays in which certain bubbles have unchangeable numbers of neighbors can be used to illustrate the thermodynamics of chemical change.

Since all vertices are trivalent, once the number of bubbles has been chosen the numbers of vertices and of their connections are also fixed for $V = 2(P - 1)$ and $E = 3(P - 1)$. Otherwise the freedom of arrangement is limited only by the boundary of the aggregate in accord with equation 3. The maximum internal freedom accompanies the minimum number of vertices on the boundary (one) and becomes zero when all vertices are associated with the boundary. Since $\Sigma nP_n + E_b$ is always $2E$, variations in the function ΣnP_n directly reflect the entropy.

This is illustrated by the simple arrays in Figure 20.15. The volume of the system is E_b, and the pressure the inverse of this. A change in structure may involve the transfer of vertices to or from the boundary,

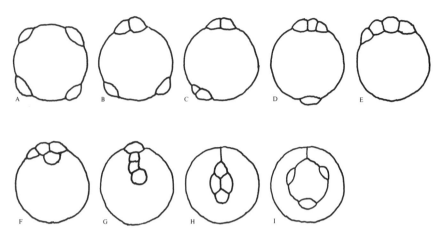

Figure 20.15. Diagrams of the dispersion and aggregation of five bubble-like polygons and their associated eight vertices to illustrate the principles of thermodynamics. (See text for details.)

which changes both volume and pressure in such a way as to keep their ratio 1, or it may be a simple internal interchange of polygon sides, as in Figures 20.15H and I, which leaves both ΣnP_n and E_b unaffected.

It immediately appears that entropy is the function $V - B$, and that $3V - \Sigma nP_n$ is analogous to temperature. If connections are counted rather than vertices, entropy is $2E/3 - E_b$ and temperature $2E - \Sigma nP_n$. The minimum "temperature", zero, comes when all vertices are on the boundary, and, conversely, the maximum temperature accompanies the arrangement with as many vertices as possible inside and a single one the boundary.

The arrays depicted in Figures 20.15A to D represent gas or liquid—note that the last three have different molecular constitution, and that C and D have different constitutions under the same temperature and pressure. Figures 20.15E to G can be regarded as either single molecules with internal structure or as solids, for there is no difference on this scale. H and I depict boundary relations from both internal and external viewpoints—also since $E_b = 1$, they represent the highest possible "temperature" with the number of vertices involved. (The pendant boundary contributes twice to the sum of polygon sides.)

Differences between the values of ΣnP_n in any two states are analogous to the change in free energy accompanying chemical combination or dissociation. If the total number of connections ($2E$ or $3V$ in our system) is taken as representing the energy in the dissociated ground state, changes in the enthalpy in accordance with the Gibbs/Helmholtz equation

$$\Delta H = \Delta G + T\Delta S \qquad (5)$$

simply reflect the fact that $2E = 3V = \Sigma n P_n$ and the total energy figured from a ground state corresponding to the physically impossible gas at 0K remains constant.

Of course, atoms and higher aggregations of electrons have changeable internal structure that is not represented in the open empty polygons of our diagrams (though it underlies the pressure differences that make the physical bubbles possible) but the variability is only in environmental cells surrounding the underlying constancies at the inner quantum levels of the atom. The behavior of such groups in relation to each other and to an outer boundary is suggested on a limited scale by the patterns of Figures 20.16 and 20.17.

Figure 20.16A shows six simple atom-like aggregates of polygons using only trivalent vertices and having inviolate trigons in the center surrounded by epidermal cells that are mainly four-sided but with some additional lines on the perimeter to allow for later joining. Figures 20.16B–E show their aggregation within a larger boundary without change in the numbers of vertices, lines, or polygons in such a way as to represent the solid, liquid, and gas states.

In all cases $\Sigma n P_n + E_b$ remains constant. The particular choice of neighbors affects the balance between temperature and entropy. Except for those associated with the inviolate atom-core polygons, a single connection from a vertex on the boundary may either join another boundary vertex or it may extend radially inward to create an epidermal cell: Its decision affects the balance of substructure throughout the whole array, however large. Conversely, a change at a location not involving the boundary is a purely local affair, with no effect whatever beyond the five boundaries that are internal to the four cells in the immediate vicinity.

Figures 20.16B and C both represent the maximum dispersion of the 25 polygons and 48 vertices in A, but the former has the smallest, the latter the largest number of vertices on the boundary. Figure 20.16C represents a region of gas at a temperature of 0K for it has the highest value of E_b possible.

Figures 20.16D and E represent stages of condensation corresponding to liquid and solid, with progressive decrease in E_b and concomitant increase in temperature associated with the heat of combination. In every case the internalization of a connection raises temperature by decreasing $\Sigma n P_n$, and this is so regardless of whether the decrease in E_b is associated with simple increase in pressure as in the gas case where the "atoms" are unchanged or whether the internalization is at another level to form a boundary interior to a "molecule".

Exactly the same relations apply to atoms having central polygons (equivalent to atomic number) of more than three sides, and if there is a second level of inviolate substructural boundaries the relations reappear

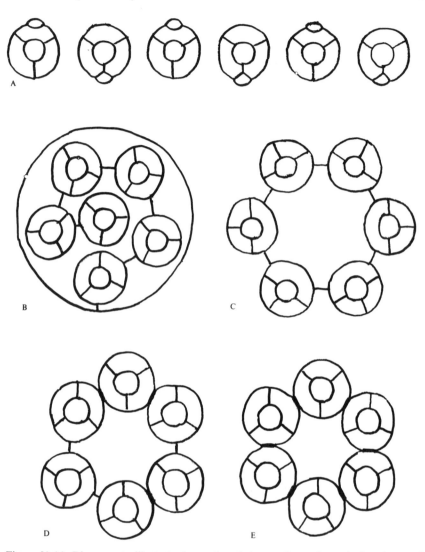

Figure 20.16. Diagrams to illustrate the analogy between thermodynamic functions and the behavior of epidermal cells surrounding an unaltered core structure. In all cases there are 48 vertices and $\Sigma nP_n + E_b$ is 144.

as those between phases as expressed by the phase rule or between differing orientations in the case of crystalline solids. Changes in entropy follow the same rules as in simpler polygons, but the ground state, by definition, excludes the core energy, for the core vertices can never lie on the boundary—though they can be oriented toward or away from it.

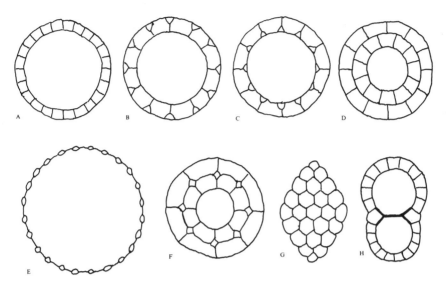

Figure 20.17. The same components as in Figure 20.16, arranged with different epidermal patterns.

Figure 20.17 shows some other combinations, all of which employ the same number of vertices (48) as in Figure 20.16, but the previously inviolate atom cores have been dissociated and rearranged to form other patterns. Figure 20.17A has the simplest epidermal structure; Figure 20.17E the maximum expansion, equivalent to minimum entropy as in Figure 20.15A, with all vertices moved to the boundary. The array in which the center is filled as uniformly as possible with hexagons, Figure 20.17G, could be converted to a conventional rectangular grid having the same number of polygons by condensing alternate pairs of trivalent vertices into single quadrivalent ones. It should be noted that it is centered on a polygon not a vertex. Figure 20.17H is interesting because of its asymmetry despite the eminent divisibility of its 48 vertices. When two aggregates having the simplest epidermal cells, as in Figure 20.17A, are joined the resulting polygons have values of $\Sigma nP_n/2$ that are not even unless $E_b - 6$ is some multiple of seventeen.

*   *   *   *

The bubble metaphor represents a combination between the atomistic and corpuscularian approaches to nature, arguments about which were a major interest of medieval philosophers. All structures, whatever their dimensionality and genus, can be resolved into trivalent nets, although when seen at lower scales of resolution they may appear to be assemblies of nets of vertices having differing valences; as polyhedra; or as the impossible objects of Maurits Escher and Roger Penrose.

It does seem that the behavior of bubbles or their representation in a two-dimensional diagram with only trivalent vertices, so easy for the human eye to comprehend, illustrates the basic principles of choice, separation, and association in general. The factors of framing, filling, and linking that Ernst Gombrich saw in decorative art provide the meeting ground between structure in human thought and that of the material universe. They are identical with the terms in Equation 5.

Postscript: Some Speculative Extensions of the Bubble Metaphor

Cosmologists have recently been finding similarities between galaxies and bubble aggregates. There are two fundamental mechanisms of change in an aggregate of soap bubbles that might be noted. Gross turbulence in a foam devolves into local shear. Although the numbers of polygons and vertices and film segments remain unchanged as bubbles slide over each other, there is an exchange of left and right neighbors. If all bubbles are the same size, and the shear is purely linear, this will cause no change in the structure, but inhomogeneous shear collects and too rapid shear creates chains of smaller bubbles with fewer than six sides. (It is undoubtedly pushing the analogy too far, but this does suggest that the particles found in the debris after high-energy nuclear collision may not be preexisting particulate components of a dense nucleus, but, rather, may be corpuscular structures *created* by the collision itself in a region of time fragmentation.)

If turbulence occurs in a normal froth, there is a tendency for smaller bubbles to segregate to the convex side of any internal boundary or the concave side of an external one. If a seemingly random net of bubbles such as those in Figure 20.4 is examined it will be found that there are long strings of associated neighbors having $n > 6$ winding around less extensive areas of smaller cells, even though the latter may be exactly hexagonal on the average (Fig. 20.18). This array was made by injecting small bubbles at one end of a flat cell and allowing them to ripen by diffusion: The initial turbulence affected the entire future history of the foam and remained after several stages of bubble growth by diffusion. A froth formed initially by growth from spatially random centers would not be expected to show such stringers, but they would develop in time as growth occurred by diffusion of gas from small high pressure bubbles to neighbors with $n > 6$.

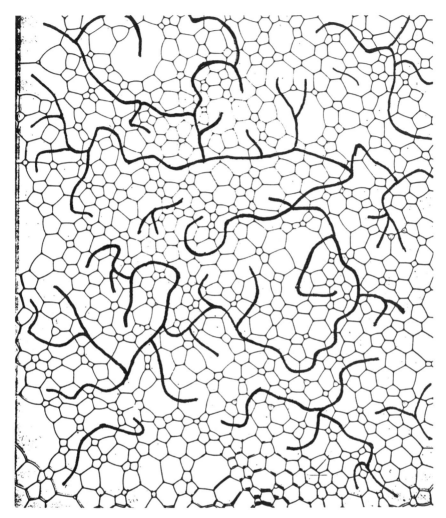

Figure 20.18. The bubble aggregate of Figure 20.3 with chains of connection between cells having more than 6 sides indicated.

Heterogeneous nucleation involves the rearrangement of cells around nuclei already present as imperfections in the preexisting structure, but homogeneous (spinodal) transformation involves the internal creation of new nuclear structures. In either case, growth from the nuclei and the eventual decrease in their number by the process known as Ostwald ripening is driven by the decrease in the number of interphase misfits and the readjustment of the hierarchy of interfaces.

In the case of bubbles, the possibility of neighborly differences disappears when the environmental pressure is high enough to force the gas phase into solution. A hole in the container would allow a turbulent jet of fine foam to escape, which in time would coarsen and segregate and condense. (Perhaps, if large enough, Taylor instability at the interface would generate Black Holes, Big Bangs, and a new evolution??)

Is it perhaps that time is the only thing that matters to matter, and The Remembrance of Things Past that is preserved in the hierarchy of the structure of matter is what we humans compare with the tiny fragments of it in our brains to produce the moiré patterns we call thought? The Oriental concept of yin/yang opposition seems to unite with the paradox of the Greek philosopher Zeno and the quantum theorist's principle of indeterminancy to suggest that there is a level of structure beneath which the pulsations of time reversal erase all substructure and make any further distinctions meaningless. The Feynman diagram (which will be seen as the lowest level of substructure in most of the illustrations in this paper) is frame, not substance. Replications of it connect to form the uniformity of space, and departures from the resulting hexagonal net become the polar and circumferential fields of electrical and magnetic interaction that, when united within a larger boundary to conform to the $(6 - n)$ rule, produce the density gradients we call gravity.